Climate **SMART** &
Energy **WISE**

This book is dedicated to learners of today and future generations as they gain the knowledge and know-how needed to survive and thrive in uncertain and turbulent times.

Climate **SMART** & Energy **WISE**

Advancing Science Literacy, Knowledge, and Know-How

Mark S. McCaffrey

Foreword by **Eugenie C. Scott** and **Jay B. Labov**

CORWIN

A SAGE Company

FOR INFORMATION:

Corwin

A SAGE Company

2455 Teller Road

Thousand Oaks, California 91320

(800) 233-9936

www.corwin.com

SAGE Publications Ltd.

1 Oliver's Yard

55 City Road

London EC1Y 1SP

United Kingdom

SAGE Publications India Pvt. Ltd.

B 1/I 1 Mohan Cooperative Industrial Area

Mathura Road, New Delhi 110 044

India

SAGE Publications Asia-Pacific Pte. Ltd.

3 Church Street

#10-04 Samsung Hub

Singapore 049483

Acquisitions Editor: Robin Najar

Associate Editor: Desirée A. Bartlett

Editorial Assistants: Ariel Price and
                                      Andrew Olson

Production Editor: Melanie Birdsall

Copy Editor: Karin Rathert

Typesetter: C&M Digitals (P) Ltd.

Proofreader: Alison Syring

Indexer: Molly Hall

Cover Designer: Karine Hovsepian

Marketing Manager: Amanda Boudria

Printed in the United States of America

Library of Congress Cataloging-in-Publication Data

McCaffrey, Mark S.
Climate smart & energy wise : advancing science literacy, knowledge, and know-how / Mark S. McCaffrey.

pages cm
Includes bibliographical references and index.

ISBN 978-1-4833-0447-2 (pbk.)

1. Climatology—Study and teaching—United States.
2. Energy consumption—Study and teaching—United States.
3. Energy conservation—Study and teaching—United States.
4. Science—Study and teaching—Curricula—United States.
5. Science—Study and teaching—Standards—United States. I. Title.

QC981.5.M44 2015
551.6076—dc23   2014017006

This book is printed on acid-free paper.

SUSTAINABLE FORESTRY INITIATIVE
Certified Chain of Custody
Promoting Sustainable Forestry
www.sfiprogram.org
SFI-01268

SFI label applies to text stock

14 15 16 17 18 10 9 8 7 6 5 4 3 2 1

Contents

Foreword

Eugenie C. Scott and Jay B. Labov

All major scientific sources—the U.S. National Academy of Sciences and the Royal Society (2014), the U.S. National Research Council (2010a, 2010b, 2010c, 2010d, 2012a, & 2013), the American Association for the Advancement of Science (http://whatweknow.aaas.org), the Intergovernmental Panel on Climate Change (2013), the National Climate Assessment (http://nca2014.globalchange.gov), multiple associations of scientists, *Science, Nature,* and other major scientific journals—concur: The best scientific information points unequivocally to recent rapid warming and a multitude of related environment perturbations of the planet. The climate has been changing since the advent of the Industrial Revolution and especially since the last half of the 20th century. Change continues unabated during this century.

What are the consequences of this climate change? Already we have seen an increase in ocean acidification and its effect on marine organisms such as corals, rising sea levels, and more extreme weather. A warmer planet has important and not yet fully understood consequences for human well-being: Where will coastal populations move as sea rise makes their habitat unsuitable? How will the population movements necessitated by sea rise affect the economy, resources, and security of surrounding nations?

Climate change results in shifting biomes, and temperature affects where and what kinds of crops can be grown. Alterations in the ranges of agricultural pests and diseases will impact human health and well-being as warmer temperatures allow the spread of insects and microorganisms previously held in check by seasonal cold. Climate change does not affect only agents of agricultural diseases; we are finding that agents of human diseases also are changing their ranges. The living things with which we share the planet similarly are experiencing and will in the future experience profound changes in habitat, which will affect their adaptations, migratory patterns, and in some cases, their survival.

The previously referenced scientific sources that agree that the planet is warming also concur that an increase in atmospheric CO_2 and other greenhouse gasses stemming from the combustion of fossil fuels have been the major cause of the post-Industrial Revolution heating of the planet. Coping with climate change is therefore linked to accommodating human energy needs, as both the numbers of people and their demands for higher living standards increase the need for more and cheaper energy. But if that additional energy continues to be derived from fossil fuels, the problem of global climate change will be exacerbated. What energy sources will we find to replace them?

During the next couple of decades as climate change and energy needs become increasingly important social and political issues, individual citizens, industries that emit large amounts of carbon into the atmosphere, and governments will need to make important economic, political, social, and ethical decisions. These decisions must be informed by and grounded in accurate and up-to-date science.

Yet the young people who are most likely to be responsible for these decisions in the near future exhibit little understanding of climate science (Leiserowitz, Smith, & Marlon, 2011). One reason for this lack of knowledge is that climate science is rarely taught comprehensively in our schools (National Research Council, 2012b). In some cases, it may not be discussed at all because teachers feel pressure from school officials, boards of education, and citizens groups to avoid addressing this and other "controversial" topics. There is evidence that in several countries outside of the United States, climate science education is assuming a more prominent place in the curriculum (Gardiner, 2014), but climate science instruction in the United States still lags behind.

The good news is that the United States may be beginning to catch up. The Next Generation Science Standards, released in 2013, call for increased coverage of climate change in middle school and high school. For those states that adopt the NGSS, climate science will become more accepted and expected as part of the science curriculum. As more climate science is taught, the number of exemplary programs and classroom resources will increase. Yet because most teachers do not encounter systematic instruction in climate science in their preservice or inservice education and many are or perceive they are restricted in what they are permitted to teach, few are prepared to adequately teach this important science.

Climate Smart & Energy Wise provides a road map to teachers to assist them in acquiring the background and resources to bring climate and energy education into their classrooms. Basic climate and energy sciences—as well as a thoughtful discussion of approaches utilized by those who deny climate change—are accompanied by pedagogical suggestions on best practices for bringing climate change into the classroom. In addition to the important task of preparing future citizens to make critically important decisions about our future, incorporating the topic of climate change into the classroom provides advantages for a harried teacher with a long checklist of expectations: It's an ideal topic to use in addressing science education standards, integrating mathematics into the science curriculum, teaching critical thinking and connections between science and society, integrating STEM (National Academy of Engineering and National Research Council, 2014), and many other responsibilities. Climate science crosscuts many educational disciplines both within and beyond the sciences and can thus promote integration across the curriculum as called for in the NGSS (http://www.nextgenscience.org) and the Common Core State Standards (http://www.corestandards.org). McCaffrey's book provides a wealth of information to help teachers find resources, including the very useful *Climate Literacy* and *Energy Literacy* frameworks, developed by scientists and master teachers. This book is packed with suggestions for where a teacher can find more information and classroom guidance for the teaching global climate change.

Educators will find inspiration in the successes of their colleagues in Chapter 6, "Programs That Work." Others have done it, so can you!

In the *Climate Literacy* Framework can be found the *Climate Literacy* "Guiding Principle," which is a simple idea with profound consequences: *Humans can take actions to reduce climate change and its impacts.* Individually and collectively, we *can* make a difference. As scientists and educators, we believe that basing those actions on a firm foundation of climate science literacy will result in better outcomes. Science cannot dictate what actions we take, but scientific evidence surely needs to inform our actions, if they are to be effective.

The *Climate Literacy* "Guiding Principle" is a simple idea that also has profound consequences for science education: Teachers can help students acquire the scientific background that will help them both understand and deal with what may be their greatest challenge as citizens. McCaffrey's book is a quick start to help teachers teach and students learn about the important topic of climate change. We hope you explore it fully.

Eugenie C. Scott, PhD, is the former Executive Director of the National Center for Science Education, Inc.

Jay B. Labov, PhD, is Senior Advisor for Education and Communication for the National Academy of Sciences and the National Research Council. Views expressed are his and not necessarily those of the NAS or the NRC.

REFERENCES

Gardiner, B. (2014, April 20). Setbacks aside, climate change is finding its way into the world's classrooms. *New York Times.* Retrieved from http://www.nytimes.com/2014/04/21/business/energy-environment/setbacks-aside-climate-change-is-finding-its-way-into-the-worlds-classrooms.html?_r=0

Intergovernmental Panel on Climate Change. (2013). *Climate change 2013: The physical science basis.* Retrieved from the World Meteorological Organization and United Nations Environment Programme at http://www.climatechange2013.org/

Leiserowitz, A., Smith, N., & Marlon, J. R. (2011). *American teens' knowledge of climate change.* A report from the Yale Project on Climate Change Communication retrieved from http://environment.yale.edu/climate-communication/files/American-Teens-Knowledge-of-Climate-Change.pdf

National Academy of Engineering and National Research Council. (2014). *STEM Integration in K–12 Education: Status, Prospects, and an Agenda for Research.* Washington, DC: National Academies Press. Retrieved from http://www.nap.edu/catalog.php?record_id=18612

National Academy of Sciences and The Royal Society. (2014). *Climate change: Evidence and causes* [5 booklets]. Washington, DC: The National Academies Press. Retrieved from http://www.nap.edu/catalog.php?record_id=18730

National Academy of Engineering and National Research Council. (2014). *STEM Integration in K–12 Education: Status, Prospects, and an Agenda for Research.* Washington, DC: National Academies Press. Retrieved from http://www.nap.edu/catalog.php?record_id=18612

National Research Council. (2010a). *Advancing the science of climate change.* Washington, DC: National Academies Press. Retrieved from http://www.nap.edu/catalog.php?record_id=12782

National Research Council. (2010b). *Adapting to the impacts of climate change.* Washington, DC: National Academies Press. Retrieved from http://www.nap.edu/catalog.php?record_id=12783

National Research Council. (2010c). *Informing an effective response to climate change.* Washington, DC: National Academies Press. Retrieved from http://www.nap.edu/catalog.php?record_id=12784

National Research Council. (2010d). *Limiting the magnitude of future climate change.* Washington, DC: National Academies Press. Retrieved from http://www.nap.edu/catalog.php?record_id=12785

National Research Council. (2012a). *Climate change: Evidence, impacts, and choices* [3 Booklets]. Washington, DC: National Academies Press. Retrieved from http://www.nap.edu/catalog.php?record_id=14674

National Research Council. (2012b). *Climate change education in formal settings, K–14: A workshop summary.* Retrieved from http://www.nap.edu/catalog.php?record_id=13435

National Research Council. (2013). *Abrupt impacts of climate change: Anticipating surprises.* Washington, DC: National Academies Press. Retrieved from http://www.nap.edu/catalog.php?record_id=18373

About the Author

Photo by Cindi Stephan

Currently serving as programs and policy director at the National Center for Science Education (NCSE), **Mark S. McCaffrey** helped spearhead the NCSE Climate Change Education initiative and convene the Climate and Energy Literacy Summit. He served as associate scientist with the Cooperative Institute for Research in Environmental Sciences (CIRES) at the University of Colorado at Boulder from 2001 to 2011. During that time he helped lead the development of the NOAA Paleo Perspective on Abrupt Climate Change and the Climate TimeLine Information Tool, and he played a catalytic role in initiating and deploying the *Climate Literacy* and *Energy Literacy* frameworks. He was a co-principle investigator of the Climate Literacy & Energy Awareness Network (CLEAN), and member of the International Polar Year Education, Outreach and Communications Committee. He holds a graduate degree in education from the University of Northern Colorado where he focused on water as an interdisciplinary and integrating theme for teaching. McCaffrey helped establish an education affinity group with the National Climate Assessment Network, which is a public-private partnership organized under the auspices of the U.S. Global Change Research Program.

He currently resides in Oakland, California, with his wife Patricia, an old Shepherd named Bella, and three cats, Mini, Leah, and Sam, in a one-bedroom apartment that's a five-minute bike ride away from his office.

Introduction

In the case of climate change, there is no Plan B because there is no Planet B.

—Christiana Figueres (2013), Executive Director,
United National Framework Convention on Climate Change

When asked if I am pessimistic or optimistic about the future, my answer is always the same: If you look at the science about what is happening on earth and aren't pessimistic, you don't understand the data. But if you meet the people who are working to restore this earth and the lives of the poor, and you aren't optimistic, you haven't got a pulse.

—Paul Hawken (2009), environmentalist, eco-entrepreneur, and author

The seeds for a new American revolution in learning—about climate, energy, and understanding and minimizing human impacts on the environment—have already been planted. Indeed, they are already starting to sprout and take root. Transforming education about climate and energy will be critical if we are to respond successfully to what may someday be seen as the greatest crisis humanity has ever faced. To succeed, this fledgling revolution needs one vital ingredient: YOU and your willingness to step up to the challenge to help transform teaching and learning for the 21st century.

EVIDENCE

This book accepts the findings of the data and peer-reviewed research of the Intergovernmental Panel on Climate Change (IPCC), the U.S. National Climate Assessment, the National Science Academies around the world, the International Energy Agency, and the U.S. Energy Information Administration. Their reports tell us the following:

- Climate change is happening.

- Human activities, especially the burning of fossil fuels and related actions, are responsible for current climate and other related global change.

- The causes and consequences are complex, significant, and serious.

- There are many things that can be done to reduce impacts and prepare for changes that are well underway.

- Energy conservation, providing energy equity and access, and moving from fossil fuels to low-carbon and renewable energy sources are important steps to reduce climate risks.

Methodically establishing and building on foundational knowledge about key concepts—weather and climate, water and carbon cycles, energy in general and in our lives in particular, and the impact of human activities on the planet—will equip learners and thereby society to make effective, informed, and evidence-based decisions.

CLARIFYING TERMS

The topics of climate and energy are brimming with technical terms and jargon, and the glossaries from the *Climate* and *Energy Literacy* frameworks are included in the appendices of this book to help define many of them. But two terms—*climate change* and *global warming*—deserve to be addressed up front.

Both are often used synonymously by the media and sometimes by scientists themselves. Both are inadequate and imprecise in ways that can lead to confusion. The term *climate change* is often used as shorthand for human-caused disruption of the climate system, but climate naturally varies widely and sometimes wildly. It is always changing, sometimes slowly, sometimes abruptly. When the term *climate change* is used, it is not always clear whether natural variability of the climate system, the explicit influence of human activities on climate, or both are being referred to. The differences are important, for first understanding naturally occurring variations and processes of change is a vital and often overlooked step to being able to accurately attribute to what extent human activities contribute to current and future changes in climate.

Global warming, which is more widely recognized and accepted by the public than *climate change*, is also problematic as a term. By emphasizing hotter temperature, it obscures the fact that a disrupted climate may in the near term trigger record low temperatures or storm events on a local or regional scale, seemingly contradicting the predicted warming. Moreover, a warmer world may seem cozy and appealing when winter winds blow, but as we will examine, sustained heating of the planet's atmosphere and surface over the coming years will result in more negative impacts than positive ones.

Recent research by Leiserowitz and colleagues (Leiserowitz, Feinberg, Rosenthal et al., 2014) has found that the term *global warming* garners higher levels of emotional engagement, support for national and personal action, and public understanding than the term *climate change*. While *climate change* can be dismissed with a glib "but climate always changes," *global warming*, flawed though it may be, does convey the reality: that the entire planet is heating. The cause: human activities.

A more accurate term for these two somewhat flawed terms is *global change*, which includes other aspects of the Earth system beyond climate. The U.S. Global Change Research Program, for instance, focuses primarily on human disruption of the climate system, but also examines other human impacts on the biosphere, such as invasive species, habitat destruction, and other environmental damage caused by human factors that may not now be directly related to increased heat in the atmosphere. As with the term *climate change, global change* also requires understanding naturally occurring processes and rates of change in order to effectively measure and understand human contributions to the observed and projected changes.

While this book does refer to climate change numerous times, the overall aim is to encourage broad climate literacy that includes an understanding of natural variability but that is particularly attuned to what the National Research Council's *K–12 Framework on Science Education* and the *Next Generation Science Standards* describe as "human impacts" on climate and, by extension, other components of the Earth system. By understanding these human impacts we can problem-solve them and better address their causes, effects, risks, and possible responses.

> This book is meant to complement and enrich environmental education efforts and the wide range of green school-related initiatives that focus on energy savings and a healthy learning environment, as well as contribute to efforts to improve science, technology, engineering, and mathematics (STEM) education.

PURPOSE

The aim of this book is to help spark conversations about climate and energy literacy and how to improve it, deepening our own and other learners' understanding and appreciation for the role of these twin topics in our lives. Most importantly, this book emphasizes that there are many things we can do to minimize these impacts and prepare for changes already underway, but we need to be informed and educated in order to be smart about addressing climate challenges and wise about energy use in our lives.

This book's purpose is not to convince anyone about the findings of climate science or make the case for a particular policy or technical solution; there are already many books that tackle those topics. Nor is this book really about conserving energy, addressing overconsumption, reducing carbon footprints, or comparing and contrasting climate and energy policy options. This book is meant to complement and enrich environmental education efforts and the wide range of green school-related initiatives that focus on energy savings and a healthy learning environment as well as contribute to efforts to improve science, technology, engineering, and mathematics (STEM) education. While the findings of climate and energy research and related implications are touched on, the emphasis here is on teaching and learning about these vital topics.

THE GOOD NEWS

There has been a convergence of factors that have been years in the making that form a near-perfect confluence of opportunities: freely available, world-class, high-quality learning resources, the revolution in mobile-learning through smart phones and tablets, new comprehensive science education standards, and in some communities the transformation of schools into living laboratories, that will allow us to make significant and rapid headway toward preparing our children and future generations for the known and unknown challenges of the future.

The Bad News

That said, we do need to acknowledge the reality of our current state of climate and energy illiteracy. On top of the impact of global change caused by human activities on the Earth system in general, we as parents, educators, leaders, citizens, and responsible human beings have done an abysmal job overall of ensuring that climate and energy are taught and taught well. We can and must do better. It is imperative that we provide students with the knowledge and critical thinking skills they need to make good energy choices and informed climate decisions throughout their lives.

More Good News

Fortunately, as a society we are beginning to take positive steps to resolve this issue of climate and energy illiteracy. Much more needs to be done, but as we will explore in depth, there are stellar educators, high-quality learning resources, educational frameworks and standards, and motivated, innovative learners that we can all benefit from and be inspired by. And while we are in many ways just beginning to build momentum and critical mass, we have begun to graduate high school and college students who have the essential climate and energy literacy—the knowledge and know-how—that they will take with them into their careers, their communities, their lives.

It is also good news that by their very complexity and far-reaching connections, climate and energy are, for practical and pedagogical reasons, ideal interdisciplinary and integrating themes in education. They touch on nearly every aspect of our lives and link with other vital issues of the 21st century, including water, economics, food, poverty, commerce, political processes, and civic engagement. Climate and energy are complementary themes. They can begin to be formally taught beginning in kindergarten and naturally bring together science and mathematics with virtually any other topic or discipline.

Indeed, climate, energy, and sustainability are already being used in education as themes for inquiry and expression in language courses, art, music, and history as well as nearly every scientific discipline. Books like *Empowering Young Voices for the Planet* (Cherry, Texley, & Lyons, 2014) showcase examples of how youth can be effectively engaged and empowered. Such efforts offer examples for how we can address the gaping hole in our current system of teaching and learning. Ultimately, these dual topics can and should be taught, in age and developmentally appropriate ways, throughout the K–12 grade levels, into higher education and/or career paths and through ongoing lifelong learning, emphasizing problem-solving and opportunities to minimize risks and maximize resilience.

Audience

While this book is primarily meant for educators to benefit the 76 million students in the United States, I hope it will be of interest to others as well: parents and grandparents who want their school-age children to have skills to confront and meet the challenges of the 21st century, administrators and policymakers, corporate leaders and decision makers, foundation officers and philanthropists, and other citizens of the planet who share the desire to do everything they can to prepare today's young people and future generations for global change.

COUNTERING CLIMATE CONFUSION AND ENERGY ERRORS

What are the key obstacles in overcoming our illiteracy? A National Research Council report (2011), *Climate Change Education: Goals, Audiences, and Strategies,* has identified some of the obstacles. In the case of climate change, the science is complex, crosscutting many disciplines, sometimes falling through curricular cracks; it hasn't been well coordinated (or funded); and the topic, while scientifically robust, can be politically and ideologically charged, which may contribute to a climate of controversy or confusion in the classroom. In the case of energy, the dynamics are similar in terms of the complexity and cross-disciplinary nature of the science, and controversy can arise when it comes to hot-button topics like nuclear power, coal mining, or hydraulic fracturing for natural gas. We will touch on these issues in this book, but the issue of controversy, real or manufactured, is important to address up front.

There is no scientific debate as to whether humans are significantly altering the climate and environmental systems, primarily through the burning of fossil fuels and associated activities that disrupt ecosystems. Despite the consensus of the scientific community on this important point, there is still substantial confusion. It is important to acknowledge that there are real debates and disagreements about various topics, such as ice sheet dynamics, the pace and scale of species extinctions, energy extraction and generation, or how rapidly sea levels will rise along vulnerable coasts. At the same time, we must be wary of myths designed to distract from the implications of current scientific findings.

NEXT GENERATION SCIENCE STANDARDS

The 2013 release of the Next Generation Science Standards (NGSS), which are "by the states, for the states" and not national standards, cover a wide range of science topics employing inquiry-based pedagogical strategies, offering enormous opportunities to incorporate climate and energy studies across grade bands. Many states and school districts have already adopted these new standards. Science teachers around the nation are planning ways to implement the concepts, practices, and core ideas in their classrooms whether or not their states are currently planning to adopt them. An essential NGSS goal is to help all students develop a solid understanding of key scientific processes and findings. The ongoing study of climate and energy fits well with this goal.

In elementary grades, for example, students learn how human activities can impact the Earth's systems, but they also discover there are ways to minimize those impacts. Through observations, they learn how weather and climate are related but are different processes, and they learn the role of energy in our lives. Building on the foundational knowledge established in elementary school, middle school students are introduced to more complex systems, including how human activities, particularly fossil fuel energy consumption and habitat destruction, are altering the climate and ecosystems. In high school, students further explore the physical and biologic processes involved with climate and energy and examine how human impacts can be minimized and appropriate technological and engineering solutions be deployed that can transform society and protect the environment.

The goals of NGSS are noble and challenging: shared standards that *all* students master measured by appropriate assessments that will ideally bolster inquiry, systems thinking, and problem-solving skills.

Educators need support to master the content themselves as well as acquire effective psychological strategies and the pedagogical tools. Appropriate professional development is imperative. Even with limited funding, we can begin to turn the tide of climate and energy illiteracy through support and leadership at the local, state, and national level, forging partnerships between public and private sectors, emphasizing teacher professional development, high-quality resources, youth empowerment, and collective impact.

GETTING STARTED

In **Chapter 1**, we will review the current scientific findings and relevant data around climate and energy science, explore the origins of the study of these and related topics, including profiling some of its pioneers, and build the case for fostering climate and energy literacy. **Chapter 2** offers some guiding principles based on education research about effective educational practices in general, focusing on science education in particular, with an emphasis on climate and energy, making the case for why these interlocking, complex topics rooted in well-established core sciences and on the frontiers of cutting-edge research should be taught throughout the curriculum. This will require leadership, collaboration, and cooperation between faculty, with the additional support of administrators and parents.

The Next Generation Science Standards (NGSS) and elements of the Common Core standards for language arts and mathematics as they relate to teaching climate and energy are examined in **Chapter 3**. When successfully deployed, NGSS will help ensure that essential climate and energy literacy, a hybrid of knowledge and know-how, is achieved by graduation.

In **Chapter 4** we will drill into the specifics of the *Climate Literacy* framework, and in **Chapter 5** we will do the same for the *Energy Literacy* framework. Examining the essential principles and fundamental concepts of each, we highlight high-quality resources across grade levels that support climate and energy science literacy. These two chapters provide practical advice and point to high-quality learning resources, including activities that can be immediately implemented by classroom educators.

Chapter 6 examines noteworthy educational programs focused on climate and energy and the elements that make them successful. **Chapter 7** is devoted to dealing with climate change skepticism and denial inside and outside of classrooms, examining the different levels of reaction, from dismissive to alarmed, and related pedagogical challenges for each group. We will delve into reasons why doubt and denial may arise and suggest when and how best to counter various forms of denial, fatalism, and apathy. In **Chapter 8**, the focus is on initiating climate and energy conversations, which have been too often missing in society, and exploring how a major initiative or public-private partnership linking national, state, and local level efforts may be an effective approach to rapidly scaling up effective climate and energy literacy practices and programs.

Finally, three appendices providing supplemental materials are included. **Appendix I**, Voices for Climate Education, is a collection of statements from a number of organizations on the imperative of teaching about climate change basics. **Appendix II** and **Appendix III** are selected excerpts of the *Climate Literacy* and *Energy Literacy* frameworks discussed in Chapters 4 and 5. These texts identify the essential principles and fundamental concepts related to climate and energy, their influence on society, and society's influence on them.

HOPE

One book cannot solve every problem. But it is hoped that readers will find this book helpful as an introduction to the topic, a practical resource for educators, and as a source of inspiration for starting the conversations and forging the partnerships necessary to create more climate-aware, energy-efficient schools, and science savvy individuals and communities. Your help in cultivating and tending the seeds that have been planted and are now sprouting will help ensure that many of these fledgling efforts can mature to provide shade and fruit for generations to come. The stakes are high, the challenges daunting. As Ana Unruh-Cohen sums up in her haiku:

As Scientists agree.

Climate change: It's real, it's us

It's bad but there's hope!

CHAPTER 1

Climate and Energy 101

HUMANS AS A FORCE OF NATURE

Since the end of the last Ice Age some 14,000 years ago, the Earth's human population has risen from at best a few million to well over seven billion, with projections of 9.6 billion by 2050 (United Nations, 2013). It is no surprise, simply by our sheer numbers, that humans have become a force of nature.

Scientists emphasize that humans now move more soil on the surface of the Earth than all natural erosion processes, that we fix more nitrogen to make fertilizers than all the nitrogen-fixing bacteria on the planet, and by altering the ocean and land cover, we have substantially altered the hydrologic and carbon cycles and impacted biodiversity.

While relatively small numbers of humans can have an oversized impact on the Earth systems, the exponential growth of population is clearly a vital piece of the overall story. But equally important is the rise in the global standard of living. Today, while extreme poverty ($1.25 a day or less) has been reduced substantially in recent years, more than one in three people on the planet live on $2.00 or less a day (World Bank, 2014). Their impact on the planet is minimal compared to those who directly or indirectly rely on substantial quantities of fossil fuels. Generally, as people emerge from poverty and basic survival mode and begin to attain some degree of affluence, their energy consumption, especially of fossil fuels, rises (see Figure 1.1).

While correlation does not always imply causation, the exponential rise of population (measured in billions of people), fossil fuel emissions (measured in billion tons of carbon dioxide per year), and human-generated carbon dioxide concentrations (measured in parts of million) are connected. So too is the measure of gross domestic product or GDP, measured in trillions of U.S. dollars, an indicator of standard of living. The numbers themselves, however, may smooth out important details, such as the fact that a relatively small percentage of the global population—an estimated one billion of the most affluent people of the planet according to one study (Chakravarty et al., 2009)—is responsible for most of the burning of fossil fuels and resulting carbon dioxide emissions, and have enjoyed the biggest benefits of increased standard of living.

To roughly measure the human impact on the climate system and environment, scientists have been using the IPAT equation (impact = population × affluence × technology) and the more climate change-specific Kaya identity equation (which boils down to

FIGURE 1.1 Fossil Fuel Emissions

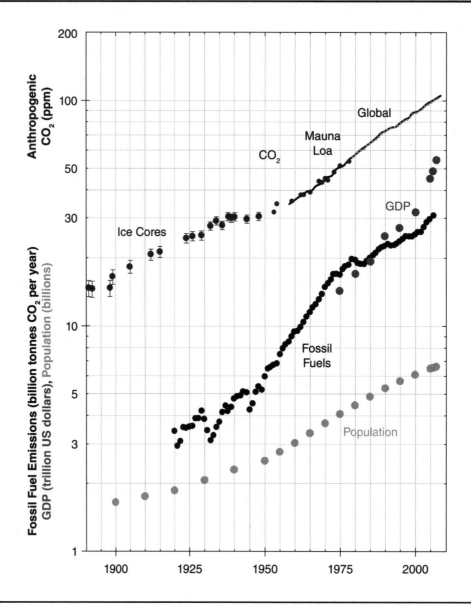

SOURCE: NOAA.

CO_2 emissions = carbon content of energy × energy consumption) (Rosa & Dietz, 2012). The result of these calculations? Carbon emissions and the use of fossil fuels continue to rise despite efforts to transition to less carbon-intensive fuels and renewable energy sources. As we will investigate throughout this book, continuing on the business-as-usual path of burning fossil fuels will lead to a substantially hotter planet with known and unknown consequences.

Yes, humans are now a geologic-scale force on the planet, with some of us having a much greater impact on the planet than others. Those who use substantial amounts of fossil fuels, including many Americans, are primarily responsible for initiating what amounts to an unintended, global-scale geoengineering experiment that is proving difficult to slow down, extremely difficult to stop or reverse.

Understanding global climate change and the connection to our energy consumption requires wrapping one's mind around its causes and effects, which is challenging, even for experts. But one way to begin to build that understanding is by focusing on the familiar elements of fire and water, light and air. In simple and complex ways, they are integral to human life

FIRE AND WATER, LIGHT AND AIR

We set fire to various fuels to cook and clean, to manufacture and burn waste. We use freshwater to grow and cook food, and the ocean as a source of protein and to dispose of wastes in. Light from the Sun and technology literally lights our lives. We live at the bottom of an ocean of air, utterly dependent on it. Reviewing some of the basics may help facilitate unpacking climate and energy in our lives and how they are intrinsically connected.

Fire or, more specifically, the oxidation of materials caused by combustion makes heat. It was the mastery of fire some 400,000 years ago that was a major breakthrough in humankind's relationship with energy. While fire produces light, what is equally attractive about fire is its heat-producing capabilities. Fire keeps people warm and cooks foods. The heat, especially when concentrated, is a source of energy, powering industry and transportation. Though often hidden away inside the dynamos of power plants or engines that propel us through space, fire generated through burning fossil fuels heats, cools, lights, and powers our lives.

It is energy from that great ball of fire known as the Sun—actually a sphere of hot plasma interwoven with magnetic fields—that drives the climate system, as well as water cycles and the process of photosynthesis. Most of the energy from the Sun that reaches Earth through the vacuum of space is in the form of shortwave visible and infrared energy traveling at the speed of light: 299,792,458 meters per second or 186,282 miles per second.

We live at the bottom of an ocean of air that we call the atmosphere. While we don't notice the pressure from the mass of air above and around us, it is an integral part of Earth's climate system. The atmosphere is mainly composed of nitrogen and oxygen. But several gases, such as carbon dioxide, water vapor, methane, and some specific artificial substances, known as greenhouse gases, are important to Earth's energy balance even though they make up a small proportion of the overall atmosphere.

Most of the incoming light energy from the Sun that penetrates through the atmosphere is shortwave infrared and visible and reaches the surface without being substantially filtered or reflected in the atmosphere. The visible light is important for photosynthetic organisms on or near the surface, which capture and store the solar energy as carbohydrates.

The shortwave incoming energy from the Sun is absorbed by Earth's surface, and some of this is then emitted as longer-wave infrared light to space; think of hot pavement or rocks radiating heat even after the Sun has set. Much of the outgoing infrared light from Earth is absorbed by greenhouse gases that capture and radiate heat, which then warms the atmosphere and by extension the rest of the planet. This process is often called the greenhouse effect, which is somewhat misleading since the warming process differs from that of an actual greenhouse.

We will investigate the Earth's climate and energy system in depth soon, but suffice to say for now that even with the direct energy from the Sun, the Earth's temperature would be roughly −18 °C, about 0 °F (Lashof, 1989), were it not for the heat radiated from greenhouse gases *after* they absorb infrared light radiated from Earth.

CLARIFYING TERMS

Heat as we normally experience and consider it—radiant heat from fireplaces, stoves, engines—occurs through the transfer of energy via convection or conduction, which cannot occur in a vacuum, such as outer space. Convection and conduction of heat through matter, primarily the air molecules of the atmosphere, play an important role in weather processes (and in heating a greenhouse), but the climate system and the greenhouse effect are driven more by thermal radiation. Thermal radiation is the warming effect that occurs when electromagnetic radiation, like the shortwave visible and infrared light from the Sun, transfers energy to atoms, causing them to vibrate faster. (The word radiation is often associated with ionizing radiation, produced by radioactivity, but in physics it is generally used to denote energy being propagated in all directions from a central source, such as the Sun.)

We normally think of light as what we can see with our human eyes—the visible spectrum from purple to red, which is between ultraviolet and infrared wavelengths. In physics, light can refer to any wavelength of electromagnetic spectrum. Thus the term *infrared light* is often preferred as more exact in terms of describing the physics than *infrared energy*, which is often translated as heat, which as described above doesn't necessarily convey thermal radiation. And finally fire as we experience it in everyday life is the combustion and oxidation of a carbon-based fuel, such as wood, dung, or a fossil fuel. The chemical processes of fire produce heat and light. But until there was sufficient free oxygen in the atmosphere from photosynthesis to allow for combustion, fire as we commonly think of it didn't exist on Earth.

Liquid water exists in relative abundance on Earth because of this natural warming effect of the planet's climate system. Water makes up much of the human body, covers most of Earth's surface, and is a vital link to understanding climate and energy in our lives. Conveniently existing on planet Earth in all three phases—solid, liquid, and gas—and even, fleetingly, all at once at the triple point of zero degrees Celsius (0.01 °C at a pressure of 611.73 pascals, to be exact)—water is in a sense the lowest common denominator for life, providing the fluid medium for organisms to exist and evolve, to survive and thrive.

The ascent of human beings from our origins on the savannahs of Africa to becoming a planetary force is often credited to our harnessing of fire, and certainly it has played an integral role in the evolution of humanity. But Peter Drucker, a business management guru, claimed that the first true technological revolution (1965) was not by fire but through the domestication of water. Drucker considered human civilization to flow from early agricultural practices involving the capture and management of water for irrigation and farming, which led in time to the development of numbers, written language, monetary systems, and city-states with armies to defend them. Were he alive today, Drucker might consider the invention of irrigation technology to be the start of what some call the *Anthropocene*, a term popularized by atmospheric chemist and Nobel laureate Paul Crutzen (Crutzen & Stoermer, 2000) to describe the current epoch or time period during which humans have become a force of geological proportions.

Certainly water, climate, and energy overlap and are often nested within each other. Today, energy to pump and clean water is required for agriculture and society. Water is needed to generate electricity, not only through hydroelectric power plants but also in most thermal electric systems, which use fossil fuels or nuclear energy to heat water into steam. The steam then turns the turbines for generating electrical power. And climate, changing seasonally through natural cycles and processes—now increasingly because of the impact from human activities—plays an integral role in how and when energy and water are consumed for human activities, whether for agricultural, industrial, or domestic purposes. Droughts not only impact agriculture but can also disrupt energy generation in power plants that require substantial amounts of water for steam generation or cooling.

BURIED SOLAR ENERGY: THE CARBON CONUNDRUM

Improved medicine, sanitation, and use of fertilizers to boost agricultural yields have made important contributions to the exponential rise of the human population and improved the quality of living in recent centuries. Education has also played an important role, generally helping counter population growth while also improving standards of living and increasing use of energy. The primary driving fuel of the global changes that humanity and the rest of the planet have been experiencing in the past several centuries, is our use of what Dukes (2003) called "buried solar energy." Fossil fuels contain energy from the Sun, originally captured by organisms through photosynthesis millions of years ago, now super concentrated through geologic time and pressure.

Fossil fuels—whether coal, natural gas, or oil—have enabled once human- and animal-powered societies, sometimes augmented by wind and water, to transform into the developed and developing civilization we live in today.

We are both indebted to and dependent on this buried solar energy. As the population has risen, so has consumption of fossil fuels, which we dig and pump from the ground to burn as fuel, harnessing the resulting heat to turn turbines and other engines to generate power. This power has deeply transformed humanity and altered the planet, including Earth's energy balance and climate system.

Appendix III, Excerpts From *Energy Literacy*, includes "A Brief History of Human Energy Use," developed primarily by physics teacher Matthew Inman for the *Energy Literacy* Framework. This is an examination of the history of humankind and our use of various forms of energy over time.

The journey of learning and teaching about climate and energy can begin with fire and water, light and air and then delve into the buried solar energy we call fossil fuels. But digging deeper requires a basic understanding of carbon and its cycle. Earth and life sciences, including biology and organic chemistry, all bring their own unique tools and perspectives to helping probe the role of carbon on the planet.

Water serves as the liquid medium, the broth if you will, for life to gestate and evolve in, but water vapor, like carbon dioxide, is a greenhouse gas molecule that absorbs infrared light and radiates heat. The two molecules also share another commonality: Both are integral to photosynthesis, the process in which organisms capture visible light from the Sun and, combining water and carbon dioxide, store that energy in a form of sugar—carbohydrates—for later use.

Photosynthesis is the base of the food chain, and over time and through geologic processes, the carbohydrates formed through photosynthesis eventually metamorphose into hydrocarbons. These can take various forms, including oil, gas, and coal, but they are all essentially buried solar energy. When they are burned, they release carbon dioxide, some water, and other gases and particulates back into the environment. What goes around comes around, as the saying goes. The process of burning fossil fuels through oxidation, where carbon locked in the fuel is released and combined with oxygen in the atmosphere, in a sense reverses the process of photosynthesis that occurred millions of years ago.

To recap, light from the Sun passes through the atmosphere, drives the hydrologic cycle, and powers ecosystems through photosynthesis. Some of the light is absorbed by the Earth, which then emits infrared light, much of which is absorbed by carbon dioxide, water vapor, and other gases that in effect trap and recirculate heat, which is often called the *greenhouse effect*. Increased concentrations of carbon dioxide from human activities (see Figure 1.2) further amplify this natural heating effect.

THE EVOLUTION OF ENERGY AND CLIMATE SCIENCE STUDY

The origins of climate science date to the 16th century. Many scientists have endeavored to understand the Earth's climate as well as the forces of energy. Some names are familiar to us, particularly those in the field of energy where the list is lengthy and the history

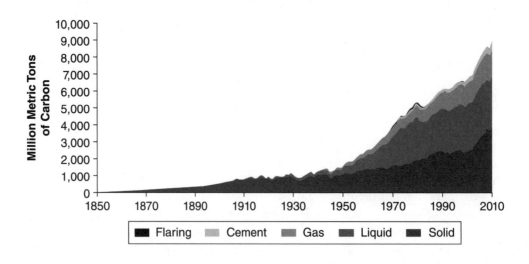

FIGURE 1.2 Global Carbon Emissions From Fossil Fuel Combustion and Cement Manufacture

SOURCE: Boden, T. A., Marland, G., & Andres, R. J. (2013). *Global, Regional, and National Fossil-Fuel CO₂ Emissions*. Retrieved from Carbon Dioxide Information Analysis Center, Oak Ridge National Laboratory, U.S. Department of Energy, Oak Ridge, TN. doi: 10.3334/CDIAC/00001_V2013

especially rich. Certainly Sir Isaac Newton (1642–1727) is one of the true giants of the field. Others include scientists and engineers who pioneered the conceptual advances and technologies that drive the modern world, including James Watt (1736–1819), Michael Faraday (1791–1867), James Joule (1818–1889), Marie Curie (1867–1934), and of course Albert Einstein (1879–1955). Along with the physicists, chemists, and geologists were industrialists—like John D. Rockefeller, George Westinghouse, and Andrew Carnegie—and inventors—such as Thomas Edison and Nikola Tesla, who transformed electrical generation, and Howard Hughes Sr., whose drill bit for oilrigs helped revolutionize the fossil fuel industry.

The names of climate scientists are less known but no less important. Information on four of the giants in climate science is detailed below. Other scientists over the years have built on the groundbreaking research of these four.

GIANTS OF CLIMATE SCIENCE

Today there are thousands of scientists around the world who contribute to our knowledge of climate science. They stand on the shoulders of four climate scientists highlighted below, who in turn benefitted from the work of Flemish chemist Jan Baptist van Helmont (1580-1644), who established that air is made up of different components and introduced the term *gas* (from the Greek word for chaos) into scientific vocabulary, and the astronomer and composer Sir William Herschel (1738-1822), who identified infrared radiation.

- Jean Baptiste Joseph Fourier (1768-1830). Well respected for his book *The Analytic Theory of Heat*, Fourier, in the 1820s, calculated that the Earth should be much cooler than it actually is given the amount of incoming energy from the Sun and the overall energy balance. He surmised that a process similar to the warming that occurs in a greenhouse may exist in the Earth's atmosphere. In fact the process is somewhat different, with a greenhouse warming through suppressed convection rather than the absorption of infrared light.
- John Tyndall (1820-1893). Born in Ireland, Tyndall was a contemporary and colleague of Charles Darwin, who was interested in the

physics of magnetism and glaciers. Knowing that most of the incoming energy from the Sun is shortwave visible and infrared radiation, and outgoing energy from the Earth is longwave infrared radiation, Tyndall sought to determine through laboratory experiments whether specific gases in the atmosphere could absorb infrared light. Through a process of elimination, he determined that the primary gases—nitrogen and oxygen—do not, but two gases in the atmosphere—water vapor and carbon dioxide—do absorb infrared wavelengths of energy, which could account for the added heat in the atmosphere that Fourier believed existed. An enthusiastic science communicator, Tyndall enjoyed developing instruments, lecturing to public audiences, and hiking in the Alps.

- Svante August Arrhenius (19 February 1859-2 October 1927). A Nobel Prize winning Swedish physicist and chemist, Arrhenius' 1896 article "On the Influence of Carbonic Acid in the Air upon the Temperature of the Ground" published in *Philosophical Magazine and Journal of Science* laid out his calculations that human activities, specifically burning coal and the resulting release of carbon dioxide (which he called

(Continued)

(Continued)

carbonic acid) into the atmosphere could increase the warming of the planet over time. He estimated that reducing carbon dioxide levels in the atmosphere by half would cause a temperature decrease of 4-5 °C (Celsius), while a doubling of carbon dioxide would result in a rise of 5-6 °C. He estimated that the doubling of carbon dioxide would take 3,000 years, but we are currently on track to have it double and perhaps triple by the year 2100, two centuries from Arrhenius' time.

- Charles David Keeling (20 April 1928-20 June 2005). As a part of the International Geophysical Year, University of San Diego scientist Charles David Keeling, working at the Scripps Institute, established a program to monitor carbon dioxide atop Hawaii's Mauna Loa in 1958. He directed the program until his death in 2005, and the project continues under the leadership of his son, Ralph F. Keeling. Their measurements allowed them to generate what is now known as the Keeling Curve, which shows the rise of carbon dioxide from less than 320 parts per million (ppm) in 1958 to about 400 ppm today. Their website (https://scripps.ucsd.edu/programs/keelingcurve) provides a detailed history of the project. NOAA also has a monitoring station on Mauna Loa, which is part of the Global Monitoring Division of the Earth System Research Laboratory, tracking data from stations around the world. The Keelings have much in common with John Tyndall, with a particular gift for developing instruments to measure carbon dioxide and its increase in the atmosphere over time.

Spencer Weart's *The Discovery of Global Warming* is a survey of the history of climate change research. Dr. Weart points out there were serious questions in the scientific community during the 19th and 20th century about the human impact on the climate system. But after the 1980s when evidence was more well established, Weart describes a shift from scientific critiques published in peer-reviewed journals to legal, political, and personal attacks in public media. "At some point they were no longer skeptics—people who would try to see every side of a case—but deniers, that is people whose only interest was in casting doubt upon what other scientists agreed was true."

A key development in climate science occurred in and around the International Geophysical Year of 1957–1958 or IGY for short. A variety of education materials were developed for IGY by the National Academy of Science, including a booklet titled *Planet Earth: The Mystery with 100,000 Clues* and educational films, such as *The Inconstant Air*. Both the booklet and film describe how burning fossil fuels add carbon dioxide to the atmosphere, adding additional heat to the atmosphere, which could lead to the melting of ice caps, the rise of sea levels, and alteration of ecosystems.

Today there are over a dozen federal agencies and thousands of scientists and support staff conducting climate and global change-related research in the United States as part of the U.S. Global Change Research Program (USGCRP). These scientists and their research make a substantial contribution to the Congressionally required National Climate Assessments of the United States and the Intergovernmental Panel on Climate Change (IPCC), which the U.S. contributes both to meet treaty obligations and as the global leader in climate research. While these reports are generally too technical for those beginning to learn the essentials of climate and energy, they are vital for those immersed in the policy implications of climate change.

FUTURE PROJECTIONS: LUKEWARM OR RED HOT?

The Intergovernmental Panel on Climate Change (IPCC) Fifth Assessment Report for Working Group II—Impacts, Adaptation, and Vulnerability (available here: http://www.ipcc-wg2.gov/index.html)—contrasts two possible scenarios based on sophisticated computer models. One is a business as usual scenario known as RCP8.5, which projects the likely outcome should the exponential growth of greenhouse gas concentrations and resulting heating of the planet occur; some parts of the world, such as the Arctic, could experience as much as 6 °C or 11 °F increased annual mean temperature above the year 2000 average by the year 2100. The other, known as RCP2.6, offers a future where carbon concentrations have been substantially flattened, which would limit global mean temperature to 2 °C, or about 3.6 °F above 2000 levels.

RCP stands for Representative Concentration Pathway (van Vuuren et al., 2011), with RCP8.5 representing a concentration of greenhouse gases that would result in the Earth retaining 8.5 watts per meter squared of radiant energy. The RCP2.6 scenario by contrast represents 2.6 watts per meter squared, only slightly higher than the current 2.3 watts per meter squared.

A group of scientists led by Detlef van Vuuren described their summary of RCP2.6 in *Climatic Change:* "Cumulative emissions of greenhouse gases from 2010 to 2100 need to be reduced by 70% compared to a baseline scenario, requiring substantial changes in energy use and emissions of non-CO_2 gases" (2011).

The RCP2.6 is considered by some to be impossible because of the massive de-carbonizing of the global economy required to achieve such levels. But the RCP8.5 is considered by many scientists to be unthinkable because of the catastrophic impact it would have on ecosystems, which are already being altered by what many scientists consider to be the sixth mass extinction (Kolbert, 2014). As Kolbert points out, most of the extinctions that are now occurring are because of human activities, such as habitat destruction and invasive species, and not directly because of heating, which thus far has been relatively modest but will increase in decades to come.

Education, carefully calibrated for the level of the learner, is clearly crucial to responding to the implications of the science and addressing the causes, effects, risks, and responses to these human-caused impacts on the climate system.

LEARNING AND TEACHING ABOUT CLIMATE SCIENCE AND ENERGY

A classic example of confusion about climate and, by extension, energy, is depicted in the Private Universe study conducted at Harvard University in the late 1980s (Macbeth, 2000). Graduating students dressed in their caps and gowns were asked a variety of questions. One was "Where does a log get its mass?" The answer that few students got: from carbon dioxide in the air, through photosynthesis. Another was "What causes the seasons?" No doubt influenced by common visualizations of the Earth's orbit around the Sun as being highly elliptical, the majority answered that the Earth gets closer to the Sun in the summer and is further from the Sun in the winter. The correct answer is, however, that the tilt of the Earth on its axis is what drives seasonal change, not distance from the Sun.

How could students at Harvard or any other university go through years of education without learning about the reason for the seasons? The reasons are simple: It's often not taught, it's not intuitive, and many teachers themselves may not have learned or understood that the Earth's 23.5-degree tilt of its axis off the perpendicular is the reason for the seasons. Two-dimensional visualizations of the Earth's orbit around the Sun often deliberately exaggerate the eccentricity of the orbit (in fact it is close to circular) and contribute to the commonly held misconception. Similarly, another key climate element with a strong energy component, the Earth's energy balance or greenhouse effect, is often not covered or taught well.

Clearly, once the basics are established, the terminology and concepts involved in understanding climate and energy become more nuanced and complex than simply fire, water, and carbon. The quantitative scales alone can boggle the mind—quads of energy, gigatons of carbon—and the technical terms are daunting, whether considering super greenhouse gases like nitrogen trifluoride or cutting edge renewable energy research like quantum photosynthesis. For both educators and learners, it is easy to feel overwhelmed by jargon and how to translate it. But each of these terms is a teachable opportunity to expand and deepen our understanding of climate and energy.

Following are a few examples of climate and energy terms that can help build vocabulary and conceptual understanding.

One quad of energy equals 10^{15} BTU, or British thermal units, which can be translated into over 8 billion gallons (U.S.) of gasoline or over 970 billion cubic feet of natural gas—enough energy to provide over 293 billion kilowatts-hours (kWh). In 2011, the U.S. produced over 78 quads of energy but consumed more than 97 quads, with 80% of the energy coming from fossil fuels, according to the U.S. Energy Information Administration (2012).

It takes one gigaton of carbon (often written at PgC for petragrams of carbon) to produce 3.67 gigatons of carbon dioxide. In 2012, according to the Global Carbon Project (2013), humans released 9.7 PgC of carbon, an increase of 2.1% from the year before, which had been the highest on record, and 58% above 1990 emissions, putting us on track for some of the most carbon intensive emission scenarios used in the Intergovernmental Panel on Climate Change (IPCC).

Nitrogen trifluoride, a human-synthesized compound used in high tech manufacturing (Robson, Gohar, Hurley, Shine, & Wallington, 2006), is an extremely long-lasting greenhouse gas that even a hundred years from now will have a global warming potential (GWP) 17,200 times greater than carbon dioxide. The significance of GWP requires its own explanation, which fortunately the U.S. EPA can help with: http://epa.gov/climatechange/ghgemissions/gases/fgases.html.

There are also many ways to link biology and other life sciences, which are not usually thought of as places to teach climate and energy, to Earth science, physics, and chemistry. For instance, quantum photosynthesis (Colini et al., 2009) involves research into understanding how quickly and efficiently plants, through quantum rather than classical physics, can transform photons of sunlight into carbohydrates. The research may lead to breakthroughs in light harvesting for generating renewable energy using marine algae.

Because climate and energy topics have often been missing or skimmed over in traditional education settings, the scale and scope of the challenges and technical jargon employed by professionals can be overwhelming. In the next several chapters we will focus on frameworks that provide learners with scaffolds that organize their understanding of these vital topics. These frameworks are based on resources developed by experts in both physical and learning sciences, and are compiled into resources that include the National Research Council's *Framework for K–12 Science Education*, the corresponding Next Generation Science Standards, and the *Climate Literacy* and *Energy Literacy* documents.

In addition to having a solid foundation in the scientific essentials, climate and energy literacy also requires an ability to discern fact from falsehood. While this book won't attempt to counter each of the extensive misconceptions and misinformation that are circulated about climate change and related topics, there are resources available to do just that. One is the Skeptical Science website (http://www.skepticalscience.com) that offers a thorough and authoritative collection of detailed analyses of climate myths and confusion that experienced science educators and learners will find useful.

THE LITERACY IMPERATIVE

The need for climate and energy literacy is urgent and compelling. Climate and global change research make it clear that human activities are pushing our planetary boundaries, and that not only are we—primarily the 20% of humanity who consume nearly 80% of the fossil fuels—heating the planet, but we are also triggering a mass extinction of other species on the planet.

Psychologically, it is not easy to accept that human-caused climate change is happening, that it is serious, that it is occurring on top of natural variations of climate, and that it can't be easily slowed or stopped. In many ways, it is easier to deny or sugarcoat what scientific findings tell us, to paint a happy face on a grim reality.

In both teaching and learning about the climate and energy challenges we face in the 21st century, we must strike a balance between pragmatic pessimism at the alarming reality and curious optimism about our ability to cope, by acknowledging the urgency and complexity of the situation but focusing on the many things we can do—at every level of society—to prepare for changes that are already well underway.

To be sure, this is nontrivial and challenging. On one hand, scientists warn that the planet is experiencing the "largest global changes in the past 65 million years" but "orders of magnitude more rapid" that raise the potential of "daunting challenges for ecosystems, especially in the context of extensive land use and degradation, changes in frequency and severity of extreme events, and interactions with other stresses" (Diffenbaugh & Field, 2013). On the other hand, there are many indicators of progress, often under the radar, as communities develop climate action and contingency plans and renewable energy, especially wind and solar, become competitive with fossil fuel electrical energy generation (Raabe, 2013).

Education is an inherently optimistic enterprise, a way to build maturity, to understand and counter risks, to foster knowledge and know-how, and to learn the whys and hows to address climate and energy challenges and opportunities. There are decisions and actions we can make today to minimize the short-term changes that will occur during our lifetimes as well as the long-term (hundreds and thousands of years) impacts that will affect future generations. But this requires that we be informed about the basic science to be better able to make decisions based on evidence rather than wishful or magical thinking.

Fostering climate and energy literacy, using fire and water and light and air as vehicles for making the connection to learners' lives, is just the first step. Literacy also requires a solid level of numeracy and data literacy, an appreciation of how current events relate to the topics, and knowledge and know-how about social, economic, and governmental processes and institutions.

Clearly, there are profound psychological, philosophical, sociological, and ethical implications and dimensions to these themes. Many aspects of climate and energy are nonintuitive, complex, and even mind-boggling. Learning (and learning to teach) about these topics is sometimes awkward and difficult. More than most subjects, the learning environment ideal for conveying these concepts should be safe but challenging, a place where all questions are good questions and everyone is able to learn from each other.

Arguably the most important quality for teaching and learning about climate and energy is a willingness to explore, to not know it all, to be open and curious, and to dive into the details in order to better grasp the massiveness of the big picture. Each of us must be comfortable cultivating our own garden, so to speak, finding a particular niche where our skills, interests, and the needs of other learners converge and we can really make a difference.

ADDITIONAL RESOURCES

DOE Energy 101 Videos, http://bit.ly/1qSZHCV

EPA Act on Climate Change Videos, http://bit.ly/1tluiL0

National Academy of Sciences & The Royal Society. (2014). *Climate Change: Evidence and Causes—An overview from the Royal Society and the U.S. National Academy of Sciences.* Retrieved February 26, 2014 from: http://dels.nas.edu/resources/static-assets/exec-office-other/climate-change-full.pdf

National Research Council (2013). *Climate change: Evidence, impacts, and choices. Answers to common questions about the science of climate change.* Retrieved February 26, 2014, from: https://nas-sites.org/ameri casclimatechoices/more-resources-on-climate-change/climate-change-lines-of-evidence-booklet

CHAPTER 2

Teaching (and Learning) About Climate Challenges and Energy Solutions

Those of us who are educators strive to develop and deploy toolkits of effective strategies based on proven results to foster the learning of our students. How do we spark learners' interest about climate and energy challenges and solutions? What will keep them engaged in the learning process? What opportunities exist for our own professional development? In this chapter we'll explore the evolution of science education regarding climate and energy science as well as ways to teach the science of climate and energy effectively.

LEARNING ABOUT LEARNING

After winning a Nobel Prize in Physics in 2001, Dr. Carl Wieman found himself on a steep learning curve as he tackled a new challenge that in many ways was as daunting and complex as his previous research on quantum physics: improving science education. He quickly discovered that funding for educational research from federal government agencies such as the National Science Foundation was limited and extremely competitive. But he persisted.

This 2004 Professor of the Year (among American doctoral and research universities) has, over the past decade, founded the PhET online catalog (http://phet.colorado. edu) of interactive simulations for physics, chemistry, biology, and Earth science courses at the University of Colorado at Boulder, where he had conducted his physics research. In 2007 he established a well-funded science education initiative at the University of British Columbia, building on his team's research on the challenges of addressing misconceptions and conveying key concepts through online tools that engage and educate.

Learning by trial and error while building on prior research about how people learn using different tools and strategies, Wieman and his team developed a suite of resources that are now used by educators and their learners around the world. They found, for instance, that virtual lab experiments for physics and chemistry could help students master concepts more effectively than actual physical labs, with the added benefit of being basically hazard and cost free.

In his role as chair of the Board on Science Education of the National Academy of Sciences, Wieman worked closely with Dr. John Holdren, President Obama's science advisor. It was while working at the White House Office of Science and Technology Policy in 2011 that he became involved with helping finalize the "Essential Principles of Energy Literacy," a document that the U.S. Department of Energy was taking the lead on developing with the input of diverse science and education experts. (We examine the principles and concepts of energy literacy in Chapter 5, and the text of the framework is available in Appendix III.)

The *Energy Literacy* project was modeled after the *Climate Literacy* "Essential Principles" led by NOAA and released in 2009, which in turn was inspired by a similar ocean literacy document on from a few years before. Like the earlier documents, the *Energy Literacy* framework provided an overview of the most vital principles and related concepts about energy in general and how it relates to our lives in particular. In addition, Wieman suggested that the *Energy Literacy* project include a guiding principle for teaching and learning.

What Wieman was suggesting went far beyond the scope of energy, addressing key elements for successful education. These universal concepts for effective teaching and learning are so relevant to setting the stage for education in general and climate and energy topics in particular, we include them below:

Guiding Principle for Teaching and Learning:

Much is understood about how people learn. Effective learning opportunities are designed with these understandings in mind.

Fundamental Concepts

1. People are born investigators and learners. People come to new learning experiences with preconceived ideas and prior knowledge. They have developed their own ideas about how the physical, biological, and social worlds work. Effective learning opportunities acknowledge and access these preconceived ideas and prior knowledge, building on correct understandings and addressing those that are incorrect.

2. Effective learning focuses on a core set of ideas and practices. Focusing on a core set of ideas and practices—rather than a broad array of what can become disconnected knowledge and isolated facts—allows a learner to make sense of new information and to tackle new problems effectively. This process is aided by explicit instructional support that stresses connections across different topics, disciplines and learning experiences.

3. Understanding develops over time. To develop a thorough understanding of the world and how it works—and to appreciate interconnections—learners need sustained opportunities, over a period of years, to work with and develop underlying ideas. People can continue learning about core ideas their entire lives. Because learning progressions extend over many years, educators must consider how topics are presented at different levels as they build on prior understanding in support of increasingly sophisticated learning.

4. Literacy requires both knowledge and practice. The social and natural sciences are not just bodies of knowledge; they are also a set of practices used to establish, extend and refine that knowledge. Effective teaching infuses these same practices into the learning experience, engaging learners in inquiry-based, authentic experiences that rely on credible information, data and evidence as the foundation for taking a position, forming conclusions or making claims.

5. Connection to interests and experiences enhances learning. Personal interest and experience is a critical part of an effective learning process. Learners must be helped to see how topics connect to their personal experience and are relevant to them. This not only aids learning in general, but also helps to foster lifelong learning.

6. Educational opportunities must be equitable and accessible to all. Effective learning requires the right tools and opportunities, and these tools and opportunities must be suited to each individual's needs.

Of all these concepts, the last is deeply embedded in public education in the United States, although the inequities and lack of opportunities are well known: Typically wealthy schools and districts attract top teachers and have the latest resources; poorer districts and schools often struggle with the basics. The educational landscape in the United States is not a level playing field, but ideally implementing the other five concepts can be achieved in any school. In fact, some of the most creative and inspiring examples of integrating climate and energy topics into the curriculum come from schools that face numerous financial and socioeconomic challenges.

But it is true that many classrooms are less than ideal learning environments. Students and even some educators may be unmotivated. Resources that support learning may be lacking. Yet even in challenging circumstances when the intentions of learners and teachers are in sync, it is awe-inspiring to witness young learners— brimming with a mixture of anxiety and excitement of the unknown, and an inherent curiosity about the ways of the world and immersing themselves in the process of making sense of the world. They must test for themselves the limits of what is possible and develop innovative, creative solutions to the often daunting and difficult challenges of life.

When we enter into a learning experience about climate or energy in a formal learning environment, we invariably bring our own preconceived ideas and prior knowledge about how physical, biological, and social systems work. Primary-grade teachers, for example, play a vital role in setting the stage for deeper learning, whether about climate, energy, or any other subject. They are instrumental in providing learning opportunities that can address the preconceived ideas, prior knowledge, and magical thinking of their young learners in order to build a clear, reality-based understanding of the world.

Learning happens inside and beyond classrooms, and increasingly it occurs online, sometimes in a computer lab but often at home or on the fly on mobile devices. One area that has potential for substantially scaling up climate and energy knowledge is the growing number of online courses for educators and learners. Some are traditional workshops or more modern hybrids that mix face-to-face workshops with online discussions. Others are MOOCs—massive open online courses—such as the Climate Literacy course run by the University of British Columbia through Coursera, Penn State's *Energy, the Environment and Our Future*, which had 28,000 students enrolled initially, or the climate leader courses offered by Climate Interactive. Such courses show potential to reach a large number of learners, including teachers learning content themselves so they can better convey it to their students.

TEACHING SCIENCE SCIENTIFICALLY

For teaching, especially the teaching of science-related topics, the goal, as Jo Handelsman and her colleagues summarize in their book *Scientific Teaching* (2007), is to make teaching more scientific, blending active learning and assessment tools and emphasizing diversity to help attain the "higher purpose" articulated by F. James Rutherford in the book *Science for All Americans* (Rutherford & Ahlgren, 1989) to "help students to develop the understanding and habits of mind they need to become compassionate human beings able to participated thoughtfully with fellow citizens in building and protecting a society that is open, decent, and vital . . ."

Achieving this noble ideal can benefit from scientific research, including such classics as *How People Learn* by the National Research Council, the legacy of constructivism dating back to John Dewey and later David Ausubel, the work on motivation and metacognition, learning styles, Bloom's taxonomy (knowledge, comprehension, application, analysis, synthesis evaluation), and the six facets of understanding identified by Wiggins and McTighe in their *Understanding by Design*: explanation, interpretation, application, perspective, empathy, self-knowledge (2005).

Handelsman and her coauthors stress the importance of active learning and assessment and sensitivity to diversity, broadly defined. While focused primarily on undergraduate education, they offer suggestions that are applicable to all educational levels, emphasizing the importance of backward design by identifying learning goals, establishing clear outcomes and assessments, and then planning learning experiences and instruction that will help attain the key goals.

Another excellent resource, although somewhat narrowly aimed at water outreach education, is the National Extension Water Outreach Education's BEPs or Best Educational Practices. Many of the suggestions are applicable to education in general and climate and energy education in particular. These resources include deconstructing the learning experience and learning goals; reflecting on the learning experience of the individual in order to best meet his or her needs; fostering environmental literacy and key principles, such as systems thinking; and identifying learners' connections to immediate surroundings in order to enhance understanding of related systems, issues, causes, and consequences.

In her 2013 essay "Teach Your Children: Ten Things the Next Generation Will Need to Know to Thrive in the Anthropocene," Minda Berbeco suggests that children need to be prepared to consider how to feed, power, hydrate, communicate, problem solve, and use diplomacy and science in order to be solutions-focused and overcome challenges. Directly or indirectly, climate and energy are in many respects ideal tools to investigate and develop skills to address these 21st century challenges.

UNITED STATES VERSUS THE WORLD

Both climate and energy are global challenges that butt up against national needs and identities in ways that may affect the teaching of these topics. For over two decades, efforts have been made to find common ground through initiatives like the UN Framework Convention on Climate Change (UNFCCC) and calls by world leaders for nations to work together to find solutions to climate change. But words, however well intended, and scientific reports and consensus, however strong, may ultimately clash with political realities at the national and local level. Nations as well as individuals quickly engage in the blame game, with much of the world blaming the United States for climate change, since historically Americans have been responsible for most of the carbon emissions on the planet. In turn, some in the United States are blaming China, now the largest emitter of gases that amplify the natural greenhouse effect.

Whatever one's opinions are about who is most to blame for climate change, when it comes to attitudes, beliefs, and knowledge about climate and energy, how does the United States compare with other nations? The short answer is: not well. Based on a range of indicators, many nations seem to be doing better at communicating or teaching about the seriousness of climate and related issues than the United States.

Before examining the data, it is important to emphasize that the U.S. public school system, which accounts for the vast majority of the 56 million students in K–12 schools, serves a broad and diverse population. Moreover, the U.S. education system—the third largest in the world after China and India—is, by design, not controlled at the national level—unlike many nations—but at the local level.

A few nations, like Germany and Canada, with provincial control of education do not have national curriculum or standards, but many countries do. By contrast, in the United States a philosophy of local control of schools at the state and ultimately school district level has historically thwarted attempts to adopt national standards, let alone a national curriculum. The Common Core State Standards and the Next Generation Science Standards are in theory "for the states, by the states," but they have been opposed in some communities because they are seen as threats to local control or because they include climate change, as in the case of Wyoming (National Center for Science Education, 2014).

Given that climate and energy have not traditionally been well covered in standard curricula, it is no surprise the United States lags behind many nations in teaching climate and energy basics. In truth, with no national standards that address all students and no systematic examination of what local school systems are doing, we do not have a comprehensive picture on how climate and energy are being taught or even if they are being taught.

There are, however, a number of important studies on international attitudes about climate and environmental issues that provide us with some clues about whether or not climate and energy instruction is occurring. Among them are the following:

- The National Geographic Society's 2012 Greendex, which ranks 17 nations in terms of the sustainable behavior of consumers. The Greendex found the United States last in terms of feeling guilt about the impact we have on the environment. But Americans are near the top of those believing individual choices make a difference. Interestingly (and perhaps counterintuitively), nations that have the least confidence that individual actions can help the environment—respectively India, China, and Brazil—lead in terms of sustainable consumer choices.

- A 2009 Pew study found that "concern about global warming is low among the publics of some big polluters—including the U.S., Russia, and China. Only about four-in-ten in the U.S. (44%) and Russia (44%) say that global warming is a very serious problem. The Chinese express the least concern—only 30% say it is a very serious problem" (2009).

- The 2009 Gallup Poll *Top-Emitting Countries Differ on Climate Change Threat* found that public awareness of climate change (using 2007–2008 data) varied from very high to very low among top emitters (Japan—99%, United States—97%, Russia—85%, China—62%, India—35%). The percentage of respondents who were both aware and considered climate change a serious personal threat ranged from 80% in Japan to 21% in China. The poll found that climate consciousness related to educational attainment, with 98% of those in China who completed four or more years of college aware of climate change and 72% of those in India. A follow-up study in 2011 found that 55% of Americans aware of climate change viewed it as a serious personal threat, down from 64% in the previous study.

- In the Gallup 2011 study *Worldwide, Blame for Climate Change Falls on Humans: Americans Among Least Likely to Attribute to Human Causes*, attribution of climate change from human activities (as opposed to natural causes or both) varied widely. Eighty-two percent of Japanese polled believed that rising temperatures are a result of human activities while only 35% of Americans embraced that belief. Only a handful of nations were lower, including Israel, Syria, Chad, and Uganda. Knowing whether and how climate change is taught in education systems may well be key to understanding these results.

- A 2013 study of college students by Eric Jamelske and colleagues titled "Comparing Climate Change Awareness, Perceptions, and Beliefs of College Students in the United States and China" in the *Journal of Environmental Sciences and Studies* found that, in comparison to their Chinese counterparts, U.S. students are significantly less likely to believe human-caused climate change is happening and were less convinced of the consensus among climate scientists regarding human-induced climate change. U.S. students regarded the economy as more important than the environment, while Chinese students ranked the economy and environment as equally important. Among both groups, who represent the future leaders of the two nations of the world that are the largest emitters of carbon dioxide with the largest economies, there was support for an international agreement to address climate change, although Chinese students supported this more strongly.

In the United States, there are schools and programs that are exemplary and innovative in their inclusion of climate and energy education in their curriculum. The Green Schools Alliance is a global network of schools working together to solve climate and conservation challenges and share sustainable best practices. The U.S. Department of Education's Green Ribbon Schools awards recognize schools "that are exemplary in reducing environmental impact and costs; improving the health and wellness of students and staff; and providing effective environmental and sustainability education" (U.S. Department of Education, 2014). A state program similar to the national program is the Texan Green Ribbon School, which emphasizes healthy diet. We will be examining some of the other stellar schools and programs that are most effective and why they are successful in Chapter 6.

Schools and programs like these demonstrate that with leadership and some basic resources, climate and energy can become integrating, interdisciplinary themes providing students with vital 21st century skills.

TEACHING CLIMATE AND ENERGY ACROSS THE CURRICULUM

When scientist Mike MacCracken started studying the physics of global climate change at the Lawrence Livermore National Laboratory (LLNL) in California in the 1980s, he teamed up with fellow scientist Manuel Perry to develop educational materials to help infuse the topic throughout the curriculum. Perry had been trained as a biochemist but also enjoyed education and public outreach for LLNL. With MacCracken's support, Perry brought together a cadre of educators from a wide range of subjects—from science to art to physical education—to develop and then test the materials in their classrooms. Every summer for five years a dozen or so master teachers would meet at the laboratory to learn directly from scientists the causes, effects, risks, and responses to climate change.

The topics covered were broad, including not only the science of global climate change, but related mathematics, social studies, language arts, geography, health, and physical education, plus integrated courses that linked them all together. In addition, the program was multicultural and multilingual, using controlled vocabulary and what today is known as sheltered English, whereby simple English with visualization is used to help English second-language learners understand the concepts.

As a pilot project, the program was a success, providing teachers with an understanding of the purpose for teaching the topic, the connections and opportunities, and a range of activities for students to become empowered by what they learned. Before funding ran out in the early 1990s, the program reached 200 teachers, with schools in Oakland, Richmond, Livermore, and the Central Valley of California using the materials. The key to the program's short-lived success, which is as relevant today as it was two decades ago, was the emphasis on fostering a culture and, if you will, a climate within an educational institution of cross-disciplinary collaboration and cooperation. This in turn requires time, leadership, and willingness on the part of all concerned to put the needs of students first, doing everything possible to prepare them for the climate and energy challenges that our communities are already facing.

MAKING LINKS, CONNECTING DOTS

While the primary emphasis of this inquiry into teaching and learning is focused on science relating to climate and energy, these topics are clearly crosscutting and interdisciplinary throughout the curriculum. Here are a few examples:

Language Arts. The Next Generation Science Standards include connections to the Common Core language and mathematics standards. Educators can build vocabulary skills around climate and energy terms and use writing, reading, and rhetoric skills to communicate the complex and contemporary nature of climate and energy topics in our lives.

Social Studies. Whether looking at civics and government, cross-cultural studies, or current affairs, every day there are climate, energy, environmental, and related social and economic issues that can be fodder for discussion and debate. Social studies broadly defined provide a complementary means of framing the science, technology, engineering, and mathematics (STEM) topics. Indeed, social studies and related humanities are imperative in order to provide balance and perspective on the findings of the sciences, and the sciences are needed to illustrate the difference between discussion based on evidence and facts and discussion based on ideology, values, and perception.

History. As we saw in Chapter 1, history takes on a fresh perspective when viewed through the lens of climate and energy use. Human history and prehistory, including the migration of our ancestors, is tied to the climate of the last Ice Age and the harnessing of fire. "A Brief History of Human Energy Use," included in the *Energy Literacy* document in Appendix III, provides an overview that can be further detailed in history classes, showing learners how we arrived at our current situation.

Geography. Once a required course in secondary school but now often an elective or missing altogether from the K–12 arena, geography is inherently an integrating and interdisciplinary course of study that examines the physical features of the planet, the regions, climates, ecosystems, peoples, and nations that make up the world. There are many geography education advocates who fight the good fight to encourage the teaching of geography, most prominently the National Geographic Society's Geo-Literacy initiative and the related Network of Alliances for Geography Education. Daniel Edelson, Vice President for Education at National Geographic, defines the three key components of geo-literacy as understanding interactions (how our world works), interconnections (how the world is connected), and implications (how interactions and interconnections determine outcomes of actions). Clearly, these elements relate with and complement climate and energy literacy.

Environmental Education. In theory, environmental education should be—and often is—a natural partner to studying climate, energy, natural sciences, and sustainable practices. But at times environmental education has been marginalized in schools, falling through curricular cracks, all but ignored in high stakes testing (Saylan & Blumbstein, 2011). It has been criticized for indoctrinating young people to become environmental activists and offering overly simplistic solutions. In some instances, environmental education is primarily focused on outdoor education, de-emphasizing science education in favor of getting people outside, enjoying the great outdoors—a worthy goal in itself but a missed opportunity for introducing or exploring environmental systems through the lens of science.

As an example of the challenges environmental education has faced, in preparing for the proposed by yet-to-be-passed No Child Left Inside Act (not to be confused with the No Child Left Behind Act), which was primarily geared toward getting young people outdoors, out of classrooms, and away from television and computer screens, states were required to develop environmental education guidelines. For the sake of getting the guidelines approved and with the aim of avoiding political controversy around funding, the guidelines avoided any mention of climate change and didn't emphasize science education. Despite having sprung from grassroots environmental movements of the 1960s and '70s, environmental education, like the environmental movement itself, evolved from grassroots-up to top-down, large, national organizations and conferences over the years. The pendulum is swinging back now, with environmental education being reinvigorated with local place-based initiatives. Ideally, focusing on climate, energy, and related topics in formal and informal education can help bring environmental awareness, concern, and informed action back into our schools and our communities in a more rigorous and educationally productive way.

Mathematics. Perhaps the most vital place in the curriculum other than science where climate and energy can be taught is mathematics, from basic arithmetic to advanced calculus and everything in between. Science relies on mathematics at every step of the way to measure, calculate, track, record, forecast, and model. News relating to climate always has ample numbers to parse, like this piece in the *Washington Post* from June, 2013: "Global emissions of carbon dioxide from energy use rose 1.4 percent to 31.6 gigatons in 2012" or any analysis of kilowatts (10^3 watts), gigawatts (10^9 watts), petawatts (10^{15} watts) and how they are generated and consumed. The inherent problem-based natures of climate and energy imperatives in our lives provide an opportunity to motivate learning around mathematical numeracy skills and data literacy in order to solve the problems.

For some students, math and science as they are currently taught are not relevant to their lives and become insurmountable hurdles that may prevent them from graduating from high school. Climate and energy can also, if not well presented, be too daunting, distant, or overwhelming. But mathematics could provide a gateway for students to relate to science. As Richard Barwell writes in his 2013 paper *The Mathematical Formatting of Climate Change: Critical Mathematics Education and Post-Normal Science*:

> Mathematics is involved at every level of understanding climate change, including the description, prediction and communication of climate change. . . . I argue that critical mathematics education offers a perspective from which to conceptualize how mathematics teaching and learning might undertake this role, drawing in particular on the idea of the "formatting power" of mathematics and the importance of reflective knowing in relation to the mathematics of climate change.

EXPANDING CLIMATE AND ENERGY LITERACY THROUGHOUT THE CURRICULUM

Psychology, sociology, business, and philosophy are generally considered the realm of higher education, but achieving climate and energy literacy during the K–12 years requires infusing the insights and perspectives of these domains into the learning experience. Following are a few thoughts on the relevance of these domains to addressing climate and energy challenges.

Psychology. Once people learn the essentials of climate change and recognize the challenges to changing behavior, they often want to know "What can we do?" and "Why is it difficult to change?" The domain of psychology can provide some deep insights into the barriers to change. For example, the American Psychological Association (APA) convened a task force (Swim et al., 2009) to look into the psychological barriers that contribute to inertia and denial. Their report, titled *Psychology and Global Climate Change: Addressing a Multi-faceted Phenomenon and Set of Challenges*, identified key factors that need to be considered in order to overcome the barriers, including uncertainty about climate change, mistrust of risk messages of experts, denial about human responsibility for changing climate, undervaluing risks because of the time lags involved, the sense of a lack of control of individuals to make a difference, and habits and ingrained behaviors. Psychology itself may not be part of K–12 curriculum, but incorporating these psychological insights into the design of K–12 climate and energy teaching will make it more effective. In higher education, where a million students take psychology courses in any given year (Munsey, 2008), the psychological dimension of climate and energy offer rich opportunities for inquiry.

Sociology. Less focused on individual psychological dynamics and more interested in our collective societal dynamics, the realm of sociology has begun to provide important insights into climate and energy challenges. Among the more influential thinkers in the sociology community, Stanley Cohen and his 2001 book *States of Denial: Knowing about Atrocities and Suffering* has identified three primary forms of collective denial that play out in society: literal denial (it's not happening), interpretive denial (it's happening but not like you think), and implicatory denial—denying the implications of the science and shirking the responsibility for what is happening.

Cohen has not written about climate change denial explicitly, but other sociologists such as Kari Norgaard (2011) have examined how the psychological challenges inherent in climate change can motivate avoidance of the topic and even a willful ignoring of the problem at a societal level. While those in the physical sciences sometimes consider both psychology and sociology "soft" sciences, both are rooted in testable theories, observations, and experimentation. Their insights and tools are invaluable for taking literacy beyond the "hard" sciences and into our lives as individuals and communities.

Philosophy. In his 2011 book *A Perfect Moral Storm: The Ethical Tragedy of Climate Change*, philosopher Stephen M. Gardiner examines in details the three factors that contribute to our current conundrum. Here's how the publisher, the Oxford University Press, describes the makings of the perfect storm:

> First, the world's most affluent nations are tempted to pass on the cost of climate change to the poorer and weaker citizens of the world. Second, the present generation is tempted to pass the problem on to future generations. Third, our poor grasp of science, international justice, and the human relationship to nature helps to facilitate inaction. As a result, we are engaging in willful self-deception when the lives of future generations, the world's poor, and even the basic fabric of life on the planet are at stake.

Of the three, Gardiner considers the intergenerational aspect to be the most problematic and in many respects the biggest leverage point.

Recognizing the philosophical underpinnings of climate change denial is another tool that can be used to make climate and energy education more effective at every level.

Business. Within business schools, there is a revolution going on that is indirectly related to climate and energy issues and often more focused on social responsibility and entrepreneurship. The efforts range from small-scale start-up efforts—helping bring renewable energy or fresh water to developing parts of the world or localizing food production—to more ambitious efforts to incubate transformative and disruptive technologies or services that can change the world.

Whatever the motivating factors and regardless as to whether the triple bottom line of "people, planets, and profit" is an explicit goal, the notion of "doing well by doing good" has been percolating through the business education community for years. CDP (https://www.cdp.net), formally the Carbon Disclosure Project, an international, not-for-profit organization that provides a global system for companies and cities to "Measure, disclose, manage and share vital environmental information," offers a treasure trove of case studies and data for students to explore as they investigate the climate and energy implications of the global supply chain. Another effort, the Sustainability Consortium, is developing a network to track not only carbon emissions in products and the supply chain but also the water footprint, the toxic waste footprint, and ultimately the social footprint of a product.

If successfully deployed, this effort will provide what Daniel Goleman (2009) calls "radical transparency" that currently is missing. Soon, consumers will be able to scan a product on their cell phone and drill down into a sustainability index that will tell the true, authenticated "story of stuff." Needless to say, the educational potential for such transparency will be enormous. Rather than having students use a black box carbon calculator to get a guesstimate of what their energy footprint is, they will be able to learn about the supply chain and how consumer choices relate to natural resources and societies in other parts of the world. What is currently invisible and unseen, whether sweatshops or carbon emissions, will be held up in the sunshine for all to see. This will indeed be revolutionary, helping tell the true story of stuff and providing students with tangible connections between climate change, energy consumption, and their own lives.

Make It Relevant

The National Center for Science Education (2012) offers four key concepts for teaching about climate change and making it relevant, which also apply to energy and many other timely topics:

- Make it local.
- Make it human.
- Make it pervasive.
- Make it hopeful.

In some respects, these are common sense, and good educators use them to seek the familiar and emphasize the positive to connect with learners. But with a global topic like climate change or the universal laws of thermodynamics, it can be a challenge to find those familiar or local points of connection.

Localizing climate and energy is important for all learners and imperative for the youngest, whose worldview and geographic orientation is limited. But whether a young learner lives in the equatorial tropics or at high latitudes or elevation, there are always ample opportunities to observe and analyze weather, seasonal changes, and the flow of energy through ecosystems and our lives. Likewise, as learners begin to contemplate and develop strategies to solve global change challenges, the local region can offer a real-world context that can then be applied to more far-flung studies in other parts of the planet.

Where does the energy we consume come from? Most of it ultimately comes from the Sun, but the pathway bringing it to our lives can be complex and challenging to unravel. And when it comes to global change, whether climate or broader human impacts, the change is not only occurring in the Arctic, in Africa, or on Pacific islands: It is happening in our own backyards and is affecting our lives, wherever we are and regardless of whether we are aware of the changes or not. One source of information about regional change in the United States is the U.S. Global Change Research Program (USGCRP, http://globalchange.gov) and especially the related National Climate Assessment (NCA, http://nca2014.globalchange.gov), which is a treasure trove of images, information, and ultimately data. Designed with mobile learners in mind, people who are not tethered to the classroom or textbook and who can decide when, where, and how they want to learn, the NCA is a potential game changer, allowing learners to delve deeply into their areas of interest, whether it is the current and future impacts of climate change in their region of the country or the actions (and future job opportunities) in specific sectors, such as agriculture and energy generation.

Indeed, planting the seed in the minds of learners about potential career paths can be a way to make the information all the more relevant and compelling. While many jobs of the future don't currently exist, many students will be intrigued by the entrepreneurial opportunities open to them if they are well informed and motivated. Whether there will be more or fewer jobs for climate scientists or solar energy physicists is not as important as having climate and energy literate doctors and nurses, lawyers and business leaders, community planners and teachers, computer programmers, technicians, farmers, and politicians who can make informed choices on simple, everyday decisions as well as major life-changing questions.

Humanizing climate and energy is also vitally important since both can be (and often are) reduced to numbers and equations that strip away the human dimension, leaving deafeningly dull data without a human soul. As the NCSE website affirms: "Science is a profoundly human endeavor, and it is important for students to appreciate that it is conducted by humans, not antiseptically deposited in timeless and impersonal textbooks" (NCSE, 2012). Bringing in the human history of energy development, which is summarized in the *Energy Literacy* framework, helps set the stage for the modern day drama and dilemma that we face with respect to our reliance on fossil fuels.

Delving into the history of climate change science can help convey the length of time and depth of research that has gone into our current understanding of the climate system and how human activities are altering it.

TEACHING THE NATURE OF SCIENCE

A familiar question in the minds of many educators and learners is "how do scientists know what they know?" There are many high-quality videos available through CLEAN and other online resources that depict scientists—increasingly many of them young and female scientists like Dr. Katey Walter Anthony at the University of Alaska—who are able to clearly, confidently, and articulately describe their research and put a face on climate science study. Also, since a great deal of learning occurs outside the classroom, climate and energy topics can and should be examined and discussed in museums, science centers, parks, and public places.

Given the grim reality—that we are currently on track for nearly 10 °F increase of global mean temperature over preindustrial levels by 2100—being hopeful about climate- and energy-related topics may seem daunting, if not delusional. Sugarcoating climate change solutions to make them appear easy—"all we have to do is change from fossil fuels to renewable energy!"—may seem to be a good approach with some students, but it is ultimately dishonest; transforming society to minimize risks and maximize preparedness will require reinventing just about everything—no small task, to be sure.

Integrating science with solutions, as the Next Generation Science Standards (NGSS) challenge educators to do, is the primary tool to counter hopelessness and despair. One relentlessly optimistic topic to raise when discussing climate and energy challenges is the Sun: the source of most energy we consume on the planet, directly or indirectly. Even schools in high latitudes or cold climates can discuss how plants and animals store solar energy from the summer to survive in the winter. While younger students may not be able to master the physics of photovoltaic or concentrated solar power, they can experience themselves how to make hot water from the Sun that will retain heat long after the Sun goes down and how that heat can be used for many purposes. At more advanced grade levels, learners can experiment with and examine other substances and methods for storing energy from renewable sources.

There are other teaching tips that naturally shake out from the focus on keeping things local, human, pervasive, and hopeful, the first one being arguably the most important of all:

Keep it simple. In the rush to convey the breadth of content and weave together all the various strands and crosscuts, to meet the diverse learners' needs and to prepare them for the assessments, whether formal or more aspirational, we may forget to start and end simply. What's the most important concept or idea, and how do we know when it is clearly understood?

Other important concepts to keep in mind are the following.

Age and Developmental Appropriateness. It may be tempting to have fourth graders in the computer lab do their own carbon footprint calculation. But remember, NGSS doesn't launch into the nuances of the carbon cycle and human impacts on the climate system until middle school for a reason: Most learners are not developmentally or emotionally ready to tackle such topics until then. Not until high school will they really be able to fully appreciate that the "reason for the seasons" is the axial tilt of the Earth.

There is more than enough age and developmentally appropriate content and skill building, analysis, and systems thinking to focus on in the primary grades. Plunging into the nuances of the carbon cycle may prove distracting from mastering an understanding of, say, the water cycle. First things first. Once the water cycle is well established, it can help frame other biogeochemical cycles, including carbon, nitrogen, and phosphorus, all of which are being disrupted by human activities.

Cultural Relevance. Being attuned to the cultural background and related needs of learners is second nature to many educators. In today's diverse and evolving culture, finding cultural relevance and resonance is a key to effective teaching. Interestingly in the United States, a 2010 poll found that 50% of Latinos and 46% of Asian-Americans "personally worry a great deal about global warming," compared with 27% of whites (Hertsgaard, 2012). Another poll (Sierra Club & National Council for La Raza, 2012) found that more than 90% of Hispanics felt that global climate change is already happening or will happen in the future. Such concern among such large segments of society offers challenges and opportunities for educators.

Inquiry and Problem-Based Learning. If there were ever problems that required deep inquiry and mastery of problem solving skills, global climate change and related energy and environmental issues are such problems. But presenting the problems in manageable, doable chunks that do not trigger apathy, fatalism, and distractions is nontrivial. Initially, keep it simple is the motto and mantra. Gradually, as learners construct their own understanding and gain confidence in their skills, the problems and inquiry can become more complex and deeper.

Addressing Misconceptions. At every step of the way, learners bring naive, muddled, and often understandable but inaccurate conceptions to the learning process (McCaffrey & Buhr, 2008). Replacing the "bad" misconceptions with "good" information is in many respects at the heart of pedagogical practice. It looks like the Sun revolves around the Earth; we even talk about it rising and setting . . . and yet the Earth is spinning at 1,674.4 km per hour (about a thousand miles per hour) at the equator and is whipping around the Sun at 108,000 kilometers per hour (around 67,000 miles per hour) even though it feels like we are standing still. We know from experience and detailed cognitive studies a simple truth: We are all capable of holding a "right" answer and understanding and simultaneously holding a "wrong" answer that on an everyday level makes more sense. Climate and energy are brimming with such paradoxes and conundrums, much to the frustration of teachers everywhere.

Teaching (and learning) about climate and energy is challenging for a plethora of reasons, but when done effectively, these two closely coupled topics can be engaging and rewarding for all concerned. There are arguably no topics in the current curriculum that are more cross disciplinary and problem based. They can and should be taught and taught well but, all too often, have been missing or skimmed over in the curriculum. Overcoming this gap and oversight can be done, but fully infusing these topics into the curriculum requires commitment, creativity, and conscious effort to foster holistic, whole-systems thinking.

A major aspect of the art of teaching science lies in overcoming student apathy, fatalism, misconceptions, and distractions. Climate, especially when framed as climate change or global warming, is particularly fraught with these impediments. On one hand, separating the science from other aspects—psychological, social, and economical—is necessary in order to focus on the learners' mastery of the underlying science. To convey the complex, often nonintuitive nature of the scientific principles and concepts, it is vital to avoid being distracted by the endless issues and opinions that are inherently wrapped into climate change.

Ideally, the science of climate, energy, and related topics that fall under the umbrella of global change could and should be taught in an integrated fashion throughout the curriculum. This can and has been done within schools where not only the science faculty but the social studies, mathematics, and history teachers all discuss how to weave together a comprehensive curriculum that helps students make the cognitive connections between what they are learning in various courses. Such cross-curricular collaboration is not easy and requires leadership and follow-through. First and foremost, science teachers (and learners) must focus on the science in order to be able to have informed, knowledgeable discussions about the social and political realms. But often the door into the science is through a news item or event that brings a meaningful context to the discussion. Part of the art of education is to use teachable moments when they arise to inform, engage, and, ideally, make the insight stick.

TERMINOLOGY MATTERS

A word on climate change (and global warming)—when a group of us began developing the "Essential Principles of Climate Literacy," we deliberately did not use the terms *climate change* or *global warming*, let alone *anthropogenic global warming* (sometimes abbreviated AGW). There were several reasons for this. One was that *climate change* and *global warming* can be highly charged terms that can distract from the task at hand: understanding the underlying science. But more importantly, understanding the processes and rates of change of the climate system and being able to comprehend whether and how humans are affecting that system (and related, interconnected systems) requires starting with the basics rather than launching head first into carbon footprints or debates about cap and trade policy.

Climate science communicator Susan Joy Hassol has been guiding scientists for years to help them better communicate their research findings to the media and public. In 2011, she gave a presentation at the TriAgency Meeting at George Mason University where she listed some of the science jargon climate scientists use in their work that is confusing if not counterproductive for the public at large. Having written with her coauthor Richard Somerville about these problematic words previously in the AGU journal *EOS* (2008) and *Physics Today* (2011), Hassol discourages scientists from using these terms with the public. But educators can and should use them as teachable moments to build vocabulary and scientific literacy among their learners through word play. Here are particular terms that Hassol identified as problematic:

- **Uncertainty.** This is certainly one of the most problematic word in science communications. Talking about uncertainty, especially at the beginning of a presentation, sounds like you don't know the topic, but scientists have been trained to present the unknowns, doubts, and caveats up front. No wonder the public is confused as to whether scientists agree that humans are impacting the climate system or not when they'll often lead with statements like, "Well, we don't know if this particular storm is related to climate change or not." It may be more effective to acknowledge that "humans have become a force of nature, and because we've altered the entire climate system to some extent, all weather has directly or indirectly been influenced by human activities."

- **GHG or greenhouse gas.** The *greenhouse* was always a metaphor, not actually how certain gases trap outgoing heat, so using the term *heat trapping gas* is an improvement.

- **Aerosol.** *Small particles* is a better term since *aerosol* means hairspray to most people.

- **Enhance.** An enhanced greenhouse effect sounds like a good thing.

- **Positive.** Positive feedback sounds like a good thing when it comes from your boss, but a positive feedback in climate is not necessarily a good thing.

- **Negative.** See above.

- **Feedback.** A positive feedback loop might be better described as a vicious cycle.

- **Radiation.** *Radiation* means nuclear energy and Chernobyl or Fukushima to most people.

- **Error.** In the scientific world, having *error bars* means clearly defining the uncertainty of the data. (Susan Hassol suggests the ideal name for a bar for scientists would be the *Error Bar*.)

Others terms that Hassol and others have flagged to watch out for and that science educators can help properly define for learners include the following:

- **Anomaly.** Something unusual or unexpected.

- **Risk.** The potential of losing something of value.

- **Driver.** A primary or contributing force of a system, such as climate.

- **Forcing.** An external influence that alters a system, such as volcanic or human impacts on climate.

- **Literature.** Sounds like old dusty books for many when in fact it is primarily peer-reviewed or other authoritative journal articles.

Also, while scientists should avoid using jargon with nontechnical audiences, learners should become familiar with phrases like SST (sea surface temperature), SLR (sea level rise), and ppm (parts per million).

Hassol's final recommendations also apply to educators and communicators in general:

- Good storytelling helps.

- Let your passion show.

- Lead with what you know.

- Consider your audience—if it's fishermen, talk about fish.

- The frame of reference and context (especially cultural and social) matters.

- Just about everyone cares about their health and children (or families).

- Make messages simple and memorable. As Professor Ed Maibach reminds us, the formula is very straightforward: Messages should be clear, they should be repeated often, and the audience should trust their messengers.

- Not all impacts of climate change will be bad . . . but we have to be honest and not gloss over the fact that negative impacts will outweigh the positive.

- The more we are able to make informed choices toward reducing impacts and preparing our communities, the slower and less extreme those impacts will be.

- Crucially, it's important to stress that it's not too late to avoid the worst impacts, and acting is cheaper than not acting.

- Moreover, taking action has many other benefits, including more livable communities, improved health, and countering denial and despair.

The energy realm has its own jargon and semantic challenges, *conservation* being the classic term. As is discussed in the first concept of the Sixth *Energy Literacy* Principle, "The amount of energy used by human society depends on many factors":

> Conservation of energy has two very different meanings. There is the physical law of conservation of energy. This law says that the total amount of energy in the universe is constant. Conserving energy is also commonly used to mean the decreased use of societal energy resources. When speaking of people conserving energy, this second meaning is always intended. (*Energy Literacy*, Appendix III)

OPINIONS AND BELIEFS IN THE CLASSROOM

As someone once said, "We're all entitled to our own opinions but not our own facts." In science, the goal of objective, empirical evidence, and repeatable, verifiable experiments trumps subjective, ideological opinions or religious beliefs. But that is not to say science can answer every question, particularly those that have a moral or ethical dimension. Nor does it mean that scientists or those who have high regard for science have no opinions or are necessarily without spiritual or religious beliefs. Far from it. Scientists may often be extremely opinionated and subject to the full range of human emotions. So while science is inherently a human endeavor, its primary goal is to address questions about the natural and material universe in a systematic and objective way. How this plays out in the science classroom is vital for effectively conveying the essence, process, and content of science. To be blunt, science is ultimately about the facts and how they are arrived at. But facts have implications, and how implications are addressed cannot be done solely by science.

Climate scientist Jeffrey Kehl has found that after his presentations, which review current science and projections of rapid warming, sea level rise, and other impacts in coming decades, he often asks audience members how the information makes them feel. Rarely are people asked such questions, and doing so in a science classroom should be undertaken with care. Pivoting from the science to what some may consider touchy-feely emotional discussions can be tricky. But people, young people in particular, do need opportunities to express their feelings of anger, hopelessness, and despair. Too often educators ask students to put on a happy face, disregarding their concerns or fears, and come up with actions they can do to lower their carbon footprint (which may already be minimal). In Chapter 7, we'll look into various types of denial and avoidance as they relate to teaching and learning about climate and energy.

FOR THE SAKE OF ARGUMENT

Finally, a few words about argumentation in science classrooms. "Whether it is new theories, novel ways of collecting data or fresh interpretations of old data, argumentation is the means that scientists use to make their case for new ideas," John Osborne (2010) argues in his article "Arguing to Learn in Science: The Role of Collaborative, Critical Discourse." Understanding the practice of science requires understanding scientific argumentation, as the experts at the National Center for Science Education acknowledge (Berbeco, McCaffrey, Meikle, & Branch, 2014), but this can be tricky and there are pitfalls. One of the key issues is being selective in the topic being argued. Here are five recommendations:

1. If a controversy is presented as a *scientific* controversy, it should be a genuine scientific controversy.

2. If a scientific controversy is presented, it should be presented at a level learners are able to understand.

3. If a scientific controversy is presented, it should be a controversy that is of manageable scope.

4. The resources for each side of the controversy ought to be comparable in quality and availability.

5. If a nonscientific controversy is presented, it should be presented as a nonscientific controversy, with the relevant scientific consensus presented first and clearly distinguished from the controversy.

SOURCE: Reprinted/adapted with permission from *The Science Teacher*, a journal for high school science educators published by the National Science Teachers Association (www.nsta.org).

In addition, it is important to avoid debating-team style arguments in a science class. That approach may be appropriate for a debating club or course in rhetoric but not a science classroom, no matter how well intentioned. Debating a misconception may backfire by reinforcing the idea that, for example, natural cycles or solar variations may be as valid as explanations of climate change as the human activities that are actually responsible for current changes. Obviously, students also need the time and guidance to prepare themselves for genuine argumentation about scientific concepts.

Infusing climate, energy, water, and sustainable practices into science courses can be a challenge since traditionally they've been skimmed over or missing. Today, though, through the NGSS, these interrelated topics and cross-cutting themes can be found in the majority of NGSS performance expectations. In the next chapter, we will begin to look for ways to link them to Common Core math and language themes and standards. With leadership and discussions with other educators and learners, these topics can become the backbone of learning in school, serving to transform schools into living laboratories for learning about 21st century science and solutions and helping schools become climate-safe, energy-efficient community centers, benefitting not only the learners they directly serve but also the entire community.

ADDITIONAL RESOURCES

CLEAN. Available at http://cleanet.org

Climate Adaptation, Mitigation, e-Learning (CAMEL). (n.d.). Available at http://www.camel climatechange.org

This is a "free, comprehensive, interdisciplinary, multi media resources for educators," primarily at the undergraduate level.

DeWaters, J. (2009). *Energy literacy survey: A broad assessment of energy-related knowledge, attitudes and behaviors.* Retrieved from http://www.esf.edu/outreach/k12/solar/2011/documents/energy_survey_HS_v3.pdf

This is as assessment instrument more specifically focused on energy.

Leiserowitz, A., Smith, N., & Marlon, J. R. (2011). *American teens knowledge of climate change.* Retrieved from http://environment.yale.edu/climate-communication/article/american-teens-knowledge-of-climate-change

Shepardson, D. P., Niyogi, D., Choi, S., & Charusombat, U. (2011). *Students' conceptions about the greenhouse effect, global warming, and climate change.* Retrieved from http://www.landsurface.org/publications-protected/J98.pdf

University of California, Berkley. (n.d.). *Global systems science curriculum for grades 9–12.* From the Lawrence Hall of Science.

Student materials available online and teacher guides available free upon request from Lawrence Hall of Science: http://www.globalsystemsscience.org

CHAPTER 3

Syncing With the Standards

What do ISO 14000 and 4-ESS3–1 have in common? Both are standards.

The first is a family of standards from the International Organization for Standardization (ISO) developed to provide "practical tools for companies and organizations looking to identify and control their environmental impact and constantly improve their environmental performance," (n.d.) helping organizations minimize negative impacts on the environment, comply with relevant laws and regulations, and assist in continual improvement of their processes.

The second is a standard being used in some U.S. schools to improve science education. It is a specific performance expectation for fourth-grade students from the Next Generation Science Standards (NGSS) on the topic of energy: "Obtain and combine information to describe that energy and fuels are derived from natural resources and their uses affect the environment" (n.d.). In theory and often in practice, standards help, well, standardize things. The National Institute for Standards and Technology (NIST), for example, helps ensure there are uniform standards on everything from atomic clocks to cyber security, building materials to nanotechnology.

In the United States, developing and implementing standards for education has been a challenging and sometimes contentious issue, in part because of a long legacy of state control of education and local control of curriculum in the country. In 1996 the National Research Council published the National Science Education Standards (NSES), which were influenced by the American Association for the Advancement of Science (AAAS) Benchmarks for Science Literacy published in 1993. Climate change caused by human activities was not included in the NSES, even though in 1992 the United States and most other nations of the world signed the United Nations Framework Convention on Climate Change (UNFCCC), which in Article 6 calls on nations to develop public education and engagement programs to understand and address climate change.

During the late 1990s, states began drafting their own science education standards, and some, like California's, at least made passing mention of the human impact on the climate system. The reasons for this general lack of coverage of the topic are numerous. In the case of the NSES, the authors said there wasn't yet sufficient data to include either global warming or its human origins in the standards. The end result of the state-by-state approach has been a hodgepodge of 50 states with their own unique science standards and uneven curriculum. Climate change and related energy topics are often not taught at all or, if they are touched upon, rarely are they covered in a comprehensive or thorough manner.

There were some efforts to develop interdisciplinary climate education programs in the 1990s, such as those developed at the Department of Energy's Lawrence Livermore Labs, the U.S. EPA's Global Warming for Kids website (now A Student's Guide to Global Climate Change), and interagency education efforts by the U.S. Global Change Research Program (USGCRP). In general, though, there was a lack of awareness, leadership, and significant funding.

In 2001, President George W. Bush asked the National Academy of Sciences to provide him with a summary of climate change science, which they did in record time, issuing a report titled *Climate Change Science: An Analysis of Some Key Questions*. On June 11, 2001, President Bush declared:

> The issue of climate change respects no border. Its effects cannot be reined in by an army nor advanced by any ideology. Climate change, with its potential to impact every corner of the world, is an issue that must be addressed by the world. (Bush, 2001)

He went on to accurately describe the fundamental issue:

> There is a natural greenhouse effect that contributes to warming. Greenhouse gases trap heat, and thus warm the earth because they prevent a significant proportion of infrared radiation from escaping into space. Concentrations of greenhouse gases, especially CO_2, have increased substantially since the beginning of the industrial revolution. And the National Academy of Sciences indicates that the increase is due in large part to human activity. (Bush, 2001)

Twelve years later, President Obama, in a speech at Georgetown University (2013), reflected that the twelve warmest years in recorded history have all come in the last fifteen years, encouraging people to "educate your classmates, your colleagues, your parents, your friends," and "broaden the circle of those who are willing to stand up for our future."

During the Bush administration, the focus was on research, and substantial research was conducted in government labs, at universities, and in research centers around the world. The Intergovernmental Panel on Climate Change (IPCC), which had been established in 1988 under the auspices of the United Nations, was reviewing climate change research and releasing periodic assessment reports providing the world with clear scientific evaluations of the current state of climate change knowledge. In 2007, the same year the organization won the Nobel Peace Prize, the Fourth Assessment of IPCC was released, detailing the strong agreement of the scientific community that the planet is warming and multiple lines of evidence point to human activities as being the primary if not overwhelming cause. The Fifth Assessment Report, made up of three working group and one synthesis report, was released in several installments beginning in late 2013.

With state science standards and curriculum generally weak and teacher professional development for climate and energy lacking, it wasn't until 2008, with the development of the *Climate Literacy* framework, that the tide began to turn. A year later, Congress authorized a number of federal agencies—specifically the National Science Foundation, NASA, and NOAA—to make available funding for grants to develop climate literacy materials and programs. Many dozens of projects were initiated during this period, and the Climate Literacy Network, now the Climate Literacy & Energy Awareness Network (CLEAN), played a role in helping identify the best education resources being developed.

CLEAN began as an informal group of advocates for climate literacy inside and outside of the federal system who were intent on building community and capacity for increased funding, research, and development of climate education materials. (Full disclosure: I was a co-founder of the original Climate Literacy Network and later helped develop CLEAN, which through funding from the National Science Foundation, has become a National Science Digital Library collection of existing online resources that have been reviewed and annotated to benefit educators and learners. Some of the key resources in the CLEAN collection are highlighted in Chapter 4.)

The next major step forward was the publication of *A Framework for K–12 Science Education,* by the National Research Council (NRC) in 2012. The framework identified key science, engineering, and technology concepts, known as disciplinary core ideas (DCI), and became the authoritative reference for the Next Generation Science Standards (NGSS), released in 2013. Global climate change is one of the DCI within the framework, and energy is one of the crosscutting themes woven throughout it.

The NGSS were developed by a consortium of science education experts from 26 states, with support from the Carnegie Foundation and coordinated by Achieve.org, which had helped facilitate the development of the Common Core State Standards for mathematics and language arts. The NGSS "lay out the disciplinary core ideas (DCI), science and engineering practices and crosscutting concepts that students should master in preparation for college and careers" (NGSS, 2012) and provide performance expectations for assessments that detail what students should be able to do at the end of instruction. The standards focus more on processes and skills—how scientists know what they know and what still remains unclear or unknown—than on memorizing facts.

These new standards reflect current scientific research and, because they have been developed "by the states, for the states," it is hoped that they will be adopted and deployed by the states on their own terms rather than be considered federally required or mandated. Most importantly, they are for all students, not just the elite few. This is particularly important for climate and other Earth science topics, which are often not taught in traditional biology, chemistry, and physics courses.

Each standard rests on the framework's three dimensions—practices, crosscutting concepts, and disciplinary core ideas. According to the NRC, "The practices describe behaviors that scientists engage in as they investigate and build models and theories about the natural world and the key set of engineering practices that engineers use as they design and build models and systems. The NRC uses the term *practices* instead of a term like *skills* to emphasize that engaging in scientific investigation requires not only skill but also knowledge that is specific to each practice." The eight practices of science and engineering that the framework identifies as essential for all students to learn are the following:

1. Asking questions (for science) and defining problems (for engineering)

2. Developing and using models

3. Planning and carrying out investigations

4. Analyzing and interpreting data

5. Using mathematics and computational thinking

6. Constructing explanations (for science) and designing solutions (for engineering)

7. Engaging in arguments based on evidence

8. Obtaining, evaluating, and communicating information

SOURCE: Reprinted with permission from *A Framework for K-12 Science Education*, 2012, by the National Academy of Sciences, Courtesy of the National Academies Press, Washington, D.C.

Because climate, energy, and related global change topics are both inherently interdisciplinary but also integrative across multiple disciplines, they are ideal topics for deploying all of these practices as well as the crosscutting concepts.

Seven crosscutting concepts make up the framework's second dimension. According to the NRC, "These concepts help provide students with an organizational framework for connecting knowledge from the various disciplines into a coherent and scientifically based view of the world" (2012, p. 83). Just as these concepts are not limited to a particular discipline, they also are not limited to a particular grade level but should be common to all as students shape their understanding of science and engineering. The seven crosscutting concepts are listed below:

1. **Patterns.** Observed patterns of forms and events guide organization and classification, and they prompt questions about relationships and the factors that influence them.

2. **Cause and effect: Mechanism and explanation.** Events have causes, sometimes simple, sometimes multifaceted. A major activity of science is investigating and explaining causal relationships and the mechanisms by which they are mediated. Such mechanisms can then be tested across given contexts and used to predict and explain events in new contexts.

3. **Scale, proportion, and quantity.** In considering phenomena, it is critical to recognize what is relevant at different measures of size, time, and energy and to recognize how changes in scale, proportion, or quantity affect a system's structure or performance.

4. **Systems and system models.** Defining the system under study—specifying its boundaries and making explicit a model of that system—provides tools for understanding and testing ideas that are applicable throughout science and engineering.

5. **Energy and matter: Flows, cycles, and conservation.** Tracking fluxes of energy and matter into, out of, and within systems helps one understand the systems' possibilities and limitations.

6. **Structure and function.** The way in which an object or living thing is shaped and its substructure determine many of its properties and functions.

7. **Stability and change.** For natural and built systems alike, conditions of stability and determinants of rates of change or evolution of a system are critical elements of study.

SOURCE: Reprinted with permission from *A Framework for K-12 Science Education*, 2012, by the National Academy of Sciences, Courtesy of the National Academies Press, Washington, D.C.

Clearly, each of these concepts is highly relevant to climate and energy education; learners begin to observe weather and seasonal patterns and examine cause and effect in elementary grades, gradually becoming more sophisticated on issues of scale, proportion, and quantitative measurements as they move into secondary grades. Systems thinking and modeling as well as measuring the flux and flow of energy and matter, whether in an ecosystem, an energy infrastructure system, or in the Earth as a system, also builds on prior knowledge, moving from the simple and local to the more complex and global. By graduation, ideally, learners are able to characterize form and function of an object or system and understand the factors that impact stability or rates of change.

In many respects, the crosscutting concepts and practices of science and engineering provide an intellectual toolkit of critical thinking skills that can be applied throughout life. But the content knowledge that the framework covers is also vital.

The third dimension of the NRC Framework is the DCI and, in NGSS the corresponding performance expectations, which "have the power to focus K–12 science curriculum, instruction, and assessments on the most important aspects of science." A core idea must have at least two of the following criteria, though some may have all four:

- Have broad importance across multiple sciences or engineering disciplines or be a key organizing concept of a single discipline

- Provide a key tool for understanding or investigating more complex ideas and solving problems

- Relate to the interests and life experiences of students or be connected to societal or personal concerns that require scientific or technological knowledge

- Be teachable and learnable over multiple grades at increasing levels of depth and sophistication (NRC, 2012)

SOURCE: Reprinted with permission from *A Framework for K-12 Science Education*, 2012, by the National Academy of Sciences, Courtesy of the National Academies Press, Washington, D.C.

The following summary focuses only on the standards that in the view of this author directly relate to energy, climate, and related human impacts on the environment. You may have your own take on the performance expectations. The following analysis is presented as an introduction rather than the final or official word on where and how climate and energy should be taught. Relevant DCI from the NRC K–12 Framework for Science Education are included for each cited standard. Each standard also includes applicable connections to the Common Core State Standards, which cover language arts and mathematics. The Common Core State Standards were released in 2010 and are not science standards but complement them.

Learning the Basics: The Natural World and How Humans Fit In

ELEMENTARY STANDARDS: ESTABLISHING THE FOUNDATION

Ideally, students should begin to learn about science in general and climate and energy in particular in elementary school, but often that is not the case. In some instances, science is missing or skimmed over for a number of reasons, including the emphasis on math and reading and high stakes testing and the lack of science background and support among elementary teachers. A national survey conducted by Horizon Research in 2012 (published 2013, p. 12) found that while math is taught all or most days in elementary grades, science is taught less frequently, with most classes offering science instruction a few days a week or a few weeks a year.

Obviously, educators should strive to be comfortable and confident in the content, appreciate the diverse learning styles of their students, and find that sweet spot of just manageable difficulties between challenging and overwhelming information where learners can construct meaning and apply it to their lives. This is especially true at the elementary level, where learners are often naturally curious, easily distracted, but potentially emotionally distressed by the facts of life, such as death.

The NGSS deliberately do not introduce human-induced climate change to students in any of the performance expectations at the elementary level. Discussing climate change in elementary school should be handled with caution, perhaps most importantly because it can distract from job one in the elementary grades: building strong content knowledge and learning skills that will benefit students in future studies, allowing them to recognize the difference between weather and climate and to set the foundation for later understanding the complex interactions of the climate system and the related socioeconomic and technological context. If the topic of climate change or global warming does come up in class, teachers can certainly take advantage of the teachable moments to discuss the topic within the frame of human impacts on the environment—which is introduced in kindergarten and continued through high school.

Scaffolding Learning

Beginning in kindergarten and continuing through the next three years, learners are immersed in observing the natural world, looking at patterns and cycles, and thinking about human impacts and possible ways to minimize those impacts. Animals and ecosystems, human systems, daily and seasonal cycles, water, and energy are all woven together in an interdisciplinary manner that will serve as a foundation for more in-depth learning.

The fourth-grade emphasis on energy in particular is a series of teachable moments for inquiry into our carbon-intensive energy infrastructure. Where and how is the electricity in our school and community generated? Where and how are transportation fuels we consume generated, and what is their impact on the environment? Are there examples of renewable energy sources being used in the community? What are the many benefits of the use of energy in our lives? Using the local community as the context and presented in general terms, these investigations will help plant the seeds for more detailed research and synthesis in the years that follow.

Throughout the elementary years and on through secondary grades, water plays an interdisciplinary and integrating role that should be tapped by educators. Water, which makes up much of the human body and covers most of the surface of the Earth, serves as the third leg of the stool, so to speak, along with climate and energy. Its ordinary, ubiquitous nature yet extraordinary qualities can serve as the vehicle to explore many nuances of climate and nature. Scientists are beginning to explore the nexus of water, energy, and climate, for the three are inseparable on planet Earth. Finding creative, meaningful ways to demonstrate where they intersect and overlap can help learners connect the dots.

One example quickly cascades into a long series of interactions: Fossil fuels are burned to turn water into steam to generate electricity, water is used to cool the turbines or reactors, energy is then used to pump water for irrigation to grow and then transport food, and energy is also used to heat our homes in the cold months and cool them in the summer. Turning these sorts of taken-for-granted interactions into teachable moments about science within our modern social and environmental context is the key to preparing young people for the energy, climate, and water challenges we already face. Helping students inquire into the energy and water systems their lives are embedded in and then factoring weather and climate dynamics into the equation can help them develop mental maps of energy, climate, and water in their lives.

Just as we don't plunge students into details of genetic drift or nuclear physics in elementary school, neither should teachers attempt to ask students to measure their carbon footprint or conduct in-depth energy audits until they have some sense of why they are doing so. Labeling carbon dioxide as bad in the minds of young students, for example, can be problematic when in later years they come to realize that the molecule is also essential for photosynthesis and humans are carbon-based organisms.

For some teachers eager to enlighten young minds about the risks of climate change and the things that can be done to mitigate carbon emissions, it is tempting to have students in fourth or fifth grade calculate their carbon footprint, watch a video such as *An Inconvenient Truth,* or debate whether climate change is happening and humans are responsible. Even in secondary grades, these are often counter-productive strategies that can lead to confusion rather than clarity. Learning progressions begin with simplified concepts and then build into increasingly complex and sophisticated understanding as the cognitive capacity of learners grows. This is important in all aspects of knowledge creation among learners, especially in the realms of climate and energy.

Establishing a solid understanding of the essentials without overwhelming learners with too much detail or gloom and doom should be the goal of elementary educators. If students move into secondary grades with an understanding of the basics of weather and climate and inquiry skills, such as how to ask scientific questions or analyze basic data on weather, climate, and energy inputs and outputs, that's an enormous accomplishment.

The same holds true for energy. Instruction should start with simple observations of energy in action and then to the potential for energy prior to action. Coal or a solar panel have potential energy that needs to be activated by another action. By the time students reach middle school, they should have examined the role and flow of energy in their lives, not only in terms of the human-created world but also the natural world and ecosystems that our technological society exists within and depends upon.

There are numerous learning activities suitable for elementary students about weather, climate, water, and energy, such as the *Beyond Weather and the Water Cycle* program from Ohio State University (http://beyondweather.ehe.osu.edu) and the Department of Energy's National Renewable Energy Laboratory (http://www.nrel.gov/education/educational_resources.html).

Introducing Science and Society

Kindergarten. In kindergarten, performance expectations are designed to help learners start to formulate inquiry questions to probe the world around them. The K-ESS3 Earth and Human Activity standard (http://www.nextgenscience.org/kess3-earth-human-activity), for example, includes three performance expectations. The third, "Communicate solutions that will reduce the impact of humans on the land, water, air, and/or other living things in the local environment," includes a clarification statement: "Examples of human impact on the land could include cutting trees to produce paper and using resources to produce bottles. Examples of solutions could include reusing paper and recycling cans and bottles."

Below the performance expectations of the standards are three boxes—Science and Engineering Practices, Disciplinary Core Ideas, and Crosscutting Concepts—mapped from the NRC K–12 Framework. Beneath that are connections to other relevant DCIs, articulation of DCIs across grade levels, and then connections to Common Core State Standards, specifically English language arts, literacy, and mathematics.

The following tables summarize the performance expectations, related DCI, and links to Common Core mathematics and language arts standards.

Note: In the description of the standards below, the use of italics indicates a direct quote from NGSS. Additional text has been added to summarize key information in the standards. Underlined text is meant to emphasize key climate or energy-related concepts.

PERFORMANCE EXPECTATION	RELATED DCI	LINKS TO COMMON CORE
K-PS2 Motion and Stability: Forces and Interactions Focusing on the basic relationship between pushes and pulls on objects and how different strengths or directions influence the speed and motion of an object. Simple experiments to measure motion are appropriate.	Forces and Motion; Types of Interactions; Relationship Between Energy and Forces; Defining Engineering Problems	Ask and answer questions, measure attributes, compare objects and describe in more/less terms.
K-PS3 Energy Observe the effects of sunlight on Earth's surface in relative terms (warmer/cooler) using different types of surface. Build a structure to reduce warming effect of sunlight, such as umbrellas or tent.	Conservation of Energy and Energy Transfer	Collaborate on research and writing projects; compare objects.

(Continued)

(Continued)

PERFORMANCE EXPECTATION	RELATED DCI	LINKS TO COMMON CORE
K-LS1 From Molecules to Organisms: Structures and Processes Describe what plants and animals (including humans) need to survive based on observation of patterns; compare needs of plants and animals and the needs of different animals; understand all plants and animals need water.	Organization for Matter and Energy Flow in Organisms	Collaborate on shared research and writing projects; compare objects.
K-ESS2 Earth's System Observe local weather conditions over time and describe patterns, different times of day, different times of year. How do plants and animals (including humans) change their environment to meet their needs? (Construct an argument supported by evidence—squirrel digs in ground to hide food, and tree roots can break concrete.)	Weather and Climate; Biogeology; Human Impacts on Earth Systems	Compose opinion pieces; compose informative/explanatory texts; collaborate on shared research and writing projects; compare objects; model with mathematics; describe measurable attributes; classify objects into categories.

SOURCE: NGSS Lead States. (2013). *Next Generation Science Standards: For States, By States*. Washington, DC: The National Academies Press.

Keeping It Simple: Science and Solutions

First Grade. Examining light, beginning to study the factors involved with the survival of plants and animals, thinking about the life cycle of organisms, and looking at patterns of objects in the sky are all part of the first-grade performance expectations.

PERFORMANCE EXPECTATION	RELATED DCI	LINKS TO COMMON CORE
1-PS4 Waves and Their Applications in Technologies for Information Transfer Focus on sound and light; investigate how different objects—transparent, translucent, opaque, reflective—affect the path of a beam of light; introduce electromagnetic radiation (potential confusion between light and sound waves).	Wave Properties; Electromagnetic Radiation; Information Technologies and Instrumentation	Write informative/explanatory texts; collaborate on shared research and writing projects; recall information or gather information to answer a question; practice collaborative conversations; order objects.
1-LS1 From Molecules to Organisms: Structures and Processes Design solutions to human problems by mimicking how plants and/or animals survive, grow, and meet their needs (biomimicry).	Structure and Function; Growth and Development of Organisms; Information Processing	Ask and answer questions in text; collaborate on shared research and writing projects; compare numbers; use powers-of-ten.
1-ESS1 Earth's Place in the Universe Observe patterns of sun, moon, and stars to describe patterns; observe amount of daylight at different times of year. (Comment: This won't be very exciting at the equator where there's almost always approximately 12 hours of day and 12 hours of night! Although it's much too early for most students to understand the Earth's axial tilt as the reason for the seasons, this is an excellent place to introduce the solstices, with longest/shortest day and night, and equinoxes, where, twice a year,	The Universe and its Stars; Earth and the Solar System	Shared research and writing; recall information to answer a question; reason abstractly and quantitatively; model with mathematics; use addition and subtraction; organize, represent, and interpret data.

everywhere on the planet—whether North Pole, South Pole and everywhere in between, there are roughly equal amounts of daylight and night. One tool that teachers can use is an annual circle chart—a large sheet of paper with a circle divided into quarters. The top, at 12, represents the summer solstice, three and nine represent the equinoxes in September and March, and six represents the winter solstice. Students can begin by finding their own birthday on the chart, and over the course of the year, the class can add notes about seasonal changes and significant weather events, and they can anticipate what will happen during the months school is no longer in session (often summer, which is an important but usually overlooked time of growth in plants and changes in the water balance of ecosystems).

SOURCE: NGSS Lead States. (2013). *Next Generation Science Standards: For States, By States*. Washington, DC: The National Academies Press.

Thinking Like A Scientist

Second Grade. Performance expectations for second graders encourage them to begin planning and carrying out investigations, collecting and analyzing data, developing explanations and solutions, and communicating what they learn from the evidence.

PERFORMANCE EXPECTATION	RELATED DCI	LINKS TO COMMON CORE
2-PS1 Matter and its Interactions *Construct an argument with evidence that some changes caused by heating or cooling can be reversed and some cannot. (Reversible changes: water or butter; irreversible: cooking an egg, freezing a leaf, burning an object.)*	Structure and Properties of Matter; Chemical Reactions	Ask and answer who, what, where, when, why, and how; collaborate on shared research and writing projects; reason abstractly and quantitatively; model with mathematics; draw a picture and a bar graph.
2-LS2 Ecosystems: Interactions, Energy, and Dynamics Determine if plants need sunlight and water to grow; examine reproduction and pollination of plants.	Interdependent Relationships in Ecosystems	Shared research and writing; add drawings or visuals; reason abstractly and quantitatively.
2-LS4 Biological Evolution: Unity and Diversity Compare general types of plants and animals in different habitats.	Biodiversity and Humans	Collaborate on shared research and writing projects; reason abstractly and quantitatively; model with mathematics.
2-ESS1 Earth's Place in the Universe Observe that Earth events can occur quickly or slowly (general, not a quantitative overview of timescales).	The History of Planet Earth	Ask and answer who, what, where, when, why, and how; shared research and writing; reason abstractly and quantitatively; model with mathematics.
2-ESS2 Earth's System Compare ways to slow or prevent wind or water from altering shape of land (such as dikes and windbreaks, plants for erosion control, etc.); model shapes and kinds of land and bodies of water in an area; identify where water is found on Earth in solid and liquid forms.	The Roles of Water in Earth's Surface; Optimizing the Design Solution	Compare and contrast two texts on same topic; use a variety of digital tools; recall or gather information; reason abstractly and quantitatively; model with mathematics.

Engineering and Technology Performance Expectations are clustered together for Grades K–2.

PERFORMANCE EXPECTATION	RELATED DCI	LINKS TO COMMON CORE
K-2 ETS1 Engineering Design Identify a problem that can be solved by developing a new or improved tool or objects; develop a representation (sketch, drawing, model) of an object to solve a problem; compare strengths and weaknesses of two objects designed to solve the same problem.	Defining and Delimiting Engineering Problems; Developing Possible Solutions; Optimizing the Design Solution	Ask and answer who, what, where, when, why, and how; collaborate on shared research and writing projects; reason abstractly and quantitatively; model with mathematics.

SOURCE: NGSS Lead States. (2013). *Next Generation Science Standards: For States, By States.* Washington, DC: The National Academies Press.

Weather, Climate and Systems Thinking

Third Grade. As they start to study the wider world, seasonal change and different weather around the world, third graders—and anyone else interested in seeing the world from a different perspective—will enjoy the Earth Wind Map from Null School (http://earth.nullschool.net) and the U.S. Wind Map (http://hint.fm/wind). The performance expectations at this grade level begin to emphasize reducing weather-related hazards and designing solutions to reduce impacts.

PERFORMANCE EXPECTATION	RELATED DCI	LINKS TO COMMON CORE
3-LS2 Ecosystems: Interactions, Energy, and Dynamics *Construct an argument that some animals form groups that help members survive.*	Social Interactions and Group Behavior	Ask and answer questions; model with mathematics; number and operations in base ten.
3-LS4 Biological Evolution: Unity and Diversity Use fossil evidence to examine organisms and environments from long ago; explore cause and affect relationships among species that may help in survival in a particular habitat. Why do some organisms in a habitat thrive or not? <u>*Make a claim about the merit of a solution to a problem caused when the environment changes and the types of plants and animals that live there may change (does not include the greenhouse effect or climate change).*</u>	Ecosystem Dynamics, Functioning and Resilience; Evidence of Common Ancestry and Diversity; Adaptation; Biodiversity and Humans	Ask and answer questions; determine main idea of text; write informative/explanatory texts; reason abstractly and quantitatively; model with mathematics; use appropriate tools strategically; draw a scaled picture graph.
3-ESS2 Earth's Systems Describe typical weather conditions expected in a particular season in data tables and graphs; <u>describe climate in different regions of the world.</u>	Weather and Climate	Ask and answer questions; compare and contrast two texts; reason abstractly and quantitatively; model with mathematics; use appropriate tools strategically; draw a scaled picture graph; measure and estimate liquid volumes using grams, kilograms, and liters.

PERFORMANCE EXPECTATION	RELATED DCI	LINKS TO COMMON CORE
3-ESS3 Earth and Human Activity _Make a claim about the merit of a design solution that reduces the impacts of weather-related hazard_ (could include barriers to prevent flooding, lightning rods, and wind resistant roofs).	Natural Hazards	Write opinion pieces; conduct short research projects; reason abstractly and quantitatively; model with mathematics.

SOURCE: NGSS Lead States. (2013). _Next Generation Science Standards: For States, By States._ Washington, DC: The National Academies Press.

Getting Energy

Fourth Grade. Emphasizing energy—how it is related to motion, how it is transferred and converted from one form to another, and how it can solve problems—the performance expectations in fourth grade also examine the Earth system, the use of maps, and factors that affect the survival of plants and animals.

PERFORMANCE EXPECTATION	RELATED DCI	LINKS TO COMMON CORE
4-PS3 Energy Provide evidence that energy can be transferred from place to place. _Apply scientific ideas to design, test, and refine a device that converts energy from one form to another_ (i.e., passive solar heater converting light to heat; assessment boundary limited to converting motion energy to electric energy or use of stored energy to cause motion or produce light or sound).	Definitions of Energy; Conservation of Energy and Energy Transfer; Relationship Between Energy and Forces; Energy in Chemical Processes and Everyday Life; Defining Engineering Problems	Refer to details and examples in a text; explain events, procedures, ideas, or concepts in a text; integrate information; write informative/explanatory texts; conduct short research projects; recall or gather relevant information; draw supporting evidence from texts; solve multistep word problems with whole numbers.
4-PS4 Waves and Their Applications in Technologies for Information Transfer Describe wave amplitude and wavelength and that waves can cause objects to move using models (not including electromagnetic waves, interference effects, or quantitative models of amplitude and wavelength).	Wave Properties; Electromagnetic Radiation; Information, Technologies and Instrumentation; Optimizing the Design Solution	Refer to details and examples in a text; integrate information from two texts; add audio recordings and visual displays; model with mathematics; draw points, lines, line segments, rays, angles, and identify in two-dimensional figures.
4-ESS1 Earth's Place in the Universe _Identify evidence from patterns in rock formations and fossils in rock layers to support an explanation for changes in a landscape over time_ (assessment boundary: limited to relative time and general processes involved).	The History of Planet Earth	Conduct short research projects; recall or gather relevant information; draw evidence from texts; reason abstractly and quantitatively; model with mathematics; know relative sizes including km, g; lb, oz; l, ml; hr, min, s (second).
4-ESS2 Earth's Systems Provide evidence of the effects of weathering or the rate of erosion; _analyze and interpret data from maps to describe patterns of Earth's features._	Earth Materials and Systems; Biogeology	Interpret information visually, orally, or quantitatively; conduct short research projects; recall or gather information; reason abstractly and quantitatively; model with mathematics; know relative sizes; solve word problems using simple fractions or decimals.

(Continued)

(Continued)

PERFORMANCE EXPECTATION	RELATED DCI	LINKS TO COMMON CORE
4-ESS3 Earth and Human Activity _Obtain and combine information to describe that energy and fuels are derived from natural resources and their uses affect the environment_ (i.e., renewable energy such as wind, hydro power, or sunlight or nonrenewable energy sources such as fossil fuels and fissile materials; considering environmental effects like loss of habitat, air pollution from burning fossil fuels); _generate and compare multiple solutions to reduce the impacts of natural Earth processes on humans_ (i.e., designing resistant buildings and improving monitoring; limited to earthquakes, floods, tsunamis, and volcanic eruptions).	Natural Resources; Natural Hazards; Designing Solutions to Engineering Problems	Refer to details and examples in text; integrate information; conduct short research projects; draw evidence from texts; reason abstractly and quantitatively; model with mathematics; interpret multiplication equation.

SOURCE: NGSS Lead States. (2013). _Next Generation Science Standards: For States, By States._ Washington, DC: The National Academies Press.

Energy and Ecosystems

Fifth Grade. In fifth grade, performance expectations add deeper dimensions to studying the Earth system, drilling into distribution of water on the planet, matter and energy cycling through ecosystems and the food chain, the role of energy from the Sun and seasonal change in driving ecosystems, and the Earth within the context of the solar system.

PERFORMANCE EXPECTATION	RELATED DCI	LINKS TO COMMON CORE
5-PS1 Matter and Its Interactions _Develop a model to describe that matter is made of particles too small to be seen;_ provide evidence that the total weight of matter is conserved regardless of the type of change that occurs when heating, cooling, or mixing substances (i.e., phase changes, dissolving, and mixing that forms new substances); identify materials based on properties; conduct investigation on whether mixing two or more substances results in a new substance.	Structure and Properties of Matter; Chemical Reactions	Draw on information and demonstrate ability to answer quickly or solve problem efficiently; conduct short research project; recall or gather relevant information; reason abstractly and quantitatively; model with mathematics; explain powers-of-ten; convert measurements; understand and measure volume.
5-PS3 Energy _Use models to describe that energy in animals' food (used for body repair, growth, motion, and to maintain body warmth) was once energy from the sun_ (using diagrams and flowcharts).	Energy in Chemical Processes and Everyday Life; Organization for Matter and Energy Flow in Organisms	Draw on information; include multimedia and visual displays in presentations.

PERFORMANCE EXPECTATION	RELATED DCI	LINKS TO COMMON CORE
5-LS1 From Molecules to Organisms: Structures and Processes Support an argument that plants get the materials they need for growth chiefly from air and water. [Clarification Statement: Emphasis is on the idea that plant matter comes mostly from air and water, not from the soil.]	Organization for Matter and Energy Flow in Organisms	Quote accurately from a text; integrate information from several texts; write opinion pieces; reason abstractly and quantitatively; model with mathematics; convert measurements.
5-LS2 Ecosystems: Interactions, Energy, and Dynamics *Develop a model to describe the movement of matter among plants, animals, decomposers, and the environment* (showing how matter that is not food becomes food within a system, such as organism, ecosystem, and the Earth).	Interdependent Relationships in Ecosystems; Cycles of Matter and Energy Transfer in Ecosystems	Draw on information; include multimedia components; reason abstractly and quantitatively; model with mathematics.
5-ESS1 Earth's Place in the Universe Emphasis on distances in space and on changes in length and direction of shadows—day and night—and seasonal appearance of some stars, such as position and motion of the Earth with respect to Sun and selected stars.	The Universe and Its Stars; Earth and the Solar System	Quote accurately from text; draw on information; explain how an author uses reasons and evidence; integrate information; write opinion piece; include multimedia; reason abstractly and quantitatively; model with mathematics; explain patterns of numbers with powers of 10; represent real world and mathematical problems in a coordinate plane.
5-ESS2 Earth's Systems Develop a model using an example to describe ways the geosphere, biosphere, hydrosphere and/or atmosphere interact. [Clarification Statement: Examples could include the influence of the ocean on ecosystems, landform shape, and climate; the influence of the atmosphere on landforms and ecosystems through weather and climate; and the influence of mountain ranges on winds and clouds in the atmosphere . . . Assessment Boundary: Assessment is limited to the interactions of two systems at a time.]	Earth Materials and Systems; The Roles of Water in Earth's Surface Processes	Draw on information; recall or gather relevant information; include multimedia; reason abstractly and quantitatively; model with mathematics; represent real world and mathematical problems in a coordinate plane.
5-ESS3 Earth and Human Activity *Obtain and combine information about ways individual communities use science ideas to protect the Earth's resources and environment.*	Human Impacts on Earth Systems	Quote accurately from text; draw on information from multiple sources; integrate information; recall or gather information; draw evidence from texts; reason abstractly and quantitatively; model with mathematics.

SOURCE: NGSS Lead States. (2013). *Next Generation Science Standards: For States, By States.* Washington, DC: The National Academies Press.

Engineering and Technology Performance Expectations are clustered together for Grades 3–5.

PERFORMANCE EXPECTATION	RELATED DCI	LINKS TO COMMON CORE
3-5-ETS1 Engineering Design *Define a simple design problem reflecting a need or a want that includes specified criteria for success and constraints on materials, time, or cost; generate and compare multiple possible solutions to a problem based on how well each is likely to meet the criteria and constraints of the problem; plan and carry out fair tests in which variables are controlled and failure points are considered to identify aspects of a model or prototype that can be improved.*	Defining and Delimiting Engineering Problems; Developing Possible Solutions; Optimizing the Design Solution	Quote accurately from text; draw on information from multiple sources; integrate information; conduct short research projects; recall or gather information; draw evidence from texts; reason abstractly and quantitatively; model with mathematics; use appropriate tools; operations and algebraic thinking.

SOURCE: NGSS Lead States. (2013). *Next Generation Science Standards: For States, By States.* Washington, DC: The National Academies Press.

MIDDLE SCHOOL: DIGGING DEEPER, STARTING SYNTHESIS

Now that a foundational understanding and tools have been established, a deeper investigation of the roles of climate and energy in our lives can begin. By middle school, learners' geographical view of the world is expanding, as is their capacity to examine the complexity of the Earth system, the intricacies of the fossil fuel-intensive infrastructure upon which many aspects of their lives depend.

Traditionally, if Earth and space science is taught at all in school, it is taught in middle school or the first year of high school. Also, traditionally, Earth science has been dominated by geology: rock and mineral types, plate tectonics, and related processes. The broader universe—galaxies, star systems, solar systems, planets, comets, and the moon—may be also included. Weather on planet Earth is often included as a unit, and climate may be mentioned, but all too often human-induced climate change, if covered at all, is treated in a hesitant way or skimmed over. In addition, the water cycle, introduced in primary grades, becomes the model for other systems integral to the Earth system, including the carbon cycle.

Next Generation Science Standards do not obviously mesh with the traditional way Earth science has been taught, though they need not necessarily clash or conflict with current teaching approaches. Because NGSS are more Earth-system science heavy than existing science standards, in the coming years educators, administrators, and other experts will need to think through and discuss how to weave the new content and performance expectations into the curriculum. Educators will need to be provided with the background and tools to include the new content, whether in a formal Earth science track or integrated into other science courses.

For years there has been a perception that colleges do not accept Earth science courses to fulfill science requirements, and this misconception has contributed to a lack of progress in expanding Earth science education. In fact, a 2013 report by the American Geosciences Institute (http://tinyurl.com/opmrbrs) has found that the majority (78%) of two- and four-year institutions do in fact accept Earth science as a legitimate science credit toward college entrance acceptance. Prior to NGSS, which makes Earth and space science equal to physical and life sciences, only one state required a yearlong Earth/environmental science course to graduate from high school.

Physical science-related performance expectations in middle school are designed to develop usable knowledge about energy—how it is defined, conserved, and transferred; its relationship to forces; and its role in chemical processes and everyday life.

In life science, the performance expectations probe the movement of matter and energy in the process of photosynthesis, bringing in crosscutting concepts, including cause and effect and structure and function as the flow of energy through ecosystems are examined.

Earth and space sciences performance expectations take a big-picture approach, looking at the planet in relationship to the universe and how patterns, such as tides, eclipses, and seasons, relate to outside influences. More in-depth investigations of processes that drive weather and influence climate are considered, with an emphasis on natural resources, hazards, and feedbacks.

One performance expectation in particular—**ESS3: Earth and Human Activity**—asks, *"How is the availability of needed natural resources related to naturally occurring processes? How can natural hazards be predicted? How do human activities affect Earth systems? How do we know our global climate is changing?"*

PERFORMANCE EXPECTATION	RELATED DCI	LINKS TO COMMON CORE
MS-PS1 Matter and Its Interactions Emphasis on atomic composition of simple molecules and extended structures; chemical reactions; making synthetic materials from natural resources; predicting particle motion and temperature when thermal energy is added or removed; examining law of conservation of matter; and designing, testing, and modifying a device to release or absorb thermal energy by chemical processes.	Structure and Properties of Matter; Chemical Reactions; Definitions of Energy; Developing Possible Solutions; Optimizing the Design Solution	Cite textual evidence to support analysis; conduct detailed experiments and research projects; gather relevant information; reason abstractly and quantitatively; model with mathematics; use ratio and rate reasoning; understand positive and negative numbers; use decimals and power of 10; display numerical data in plots; summarize numerical data.
MS-PS2 Motion and Stability: Forces and Interactions Emphasis on Newton and Kepler's laws; motion and change relating to sum of forces and mass of object; and examining electric and magnetic forces.	Forces and Motion; Types of Interactions	Cite specific text; conduct experiments; write discipline-specific arguments; conduct short research projects; reason abstractly; understand positive and negative numbers; write, read, and evaluate algebraic equations; solve multi-step, real-life and mathematical problems; use variables to solve problems.
MS-PS3 Energy Emphasis on relationships of kinetic energy to mass and speed of an object; how changes in arrangement of objects alters amount of relative potential energy stored in a system; minimizing or maximizing thermal energy transfer; <u>relationship between temperature, kinetic energy; and type of matter and mass and transfer of energy through motion.</u>	Definitions of Energy; Conservation of Energy and Energy Transfer; Relationship Between Energy and Forces; Defining and Delimited an Engineering Problem; Developing Possible Solutions	Cite specific text; conduct experiments; write discipline-specific arguments; conduct short research projects; integrate multimedia and visuals; reason abstractly; understand ratios; understand unit rate; recognize proportional relationships; know and apply integer exponents; use square and cube roots; understand linear and nonlinear functions; summarize numerical data sets in context.

(Continued)

PERFORMANCE EXPECTATION	RELATED DCI	LINKS TO COMMON CORE
MS-PS4 Waves and Their Applications in Technologies for Information Transfer Emphasis on mechanical (oscillation of matter) and electromagnetic waves; understanding how waves are reflected, absorbed, or transmitted through various materials and how digitized signals (wave pulses) are encoded and transmitted.	Wave Properties; Electromagnetic Radiation; Information, Technologies and Instrumentation	Cite specific texts; determine central idea of text and summarize; compare and contrast information; draw evidence from texts; integrate multimedia and visuals; reason abstractly; model with mathematics; understand ratios; use ratio and rate reasoning; recognize proportional relationships; understand linear and nonlinear functions.
MS-LS1 From Molecules to Organisms: Structures and Processes Most relevant to climate and energy is the <u>focus on the role of photosynthesis in the cycling of matter and flow of energy into and out of organisms; how food is rearranged through chemical reactions to form new molecules to support grown and/or release energy;</u> and how environmental and genetic factors (including availability of energy) influence growth of organisms.	Structure and Function; Growth and Development of Organisms; Organization for Matter and Energy Flow in Organisms; Information Processing; Energy in Chemical Processes and Everyday Life	Cite specific text; determine central ideas; evaluate claims in text; write arguments; write informative/explanatory texts; conduct short research projects; draw evidence from texts; integrate multimedia and visuals; use variables and write equations; understand statistical data; summarize numerical data.
MS-LD2 Ecosystems, Interactions, Energy, and Dynamics Emphasis on resource availability on organisms and populations in an ecosystem and patterns of interactions among organisms across multiple ecosystems; <u>modeling cycling of matter and flow of energy among living and nonliving parts of an ecosystem</u> (including conservation of matter and flow of energy into and out of various ecosystems); examining how changes to physical or biological components of an ecosystem affect populations; and <u>evaluating design solutions to maintaining biodiversity and ecosystem services, such as water purification, nutrient recycling, prevention of soil erosion within scientific, economic, and social context.</u>	Interdependent Relationships in Ecosystems; Cycle of Matter and Energy Transfer in Ecosystems; Ecosystem Dynamics, Functioning, and Resilience; Biodiversity and Humans; Developing Possible Solutions; Information Processing; Energy in Chemical Processes and Everyday Life	Cite specific text; integrate text and visuals; distinguish among facts, reasoned judgment, and speculation in text; write arguments; write informative/explanatory texts; draw evidence from texts; engage in collaborative discussions; present claims and findings; include multimedia; model with mathematics; use ratio and rate reasoning; use variables and write equations; summarize numerical data.
MS-ESS1 Earth's Place in the Universe <u>Emphasis on Earth-sun cyclic patterns reason for the seasons;</u> scale and properties of solar system; origin of the Earth (4.6 billion years ago); geologic timescale and related major events, such as earliest evidence of life and ice ages.	The Universe and Its Stars; Earth and the Solar System; The History of Planet Earth	Cite specific text; integrate information in text with visuals; write informative/explanatory texts; include multimedia; reason abstractly and quantitatively; model with mathematics; understand and use ratio language; recognize proportional relationships, use variables and write equations.
MS-ESS2 Earth's System <u>Emphasis on cycling of Earth's materials and flow of energy; changes of Earth's surface at varying time and spatial scales; cycling of water through Earth's system driven by energy from the sun and gravity;</u> motions and complex interactions of air masses and impact of unequal heating; and the <u>Earth's rotation and oceanic circulation in determining regional climate.</u>	The History of Planet Earth; Earth's Materials and Systems; Plate Tectonics and Large-Scale System Interactions; The Roles of Water in Earth's Surface Processes; Weather and Climate	Cite specific text; integrate information in text with visuals; compare and contrast information; write informative/explanatory texts; gather information; include multimedia; reason abstractly and quantitatively; understand positive and negative numbers; use variables and write equations.

CLIMATE SMART & ENERGY WISE

PERFORMANCE EXPECTATION	RELATED DCI	LINKS TO COMMON CORE
MS-ESS3 Earth and Human Activity Emphasis on uneven distributions of minerals, energy and groundwater on the Earth; understanding and preparing for natural hazards and catastrophic events; monitoring and minimizing human impacts on the environment; analyzing impact of human population and per capita consumption of natural resources on Earth's systems; and asking questions to *clarify evidence of the factors that have caused the rise in global temperatures over the past century . . .* such as fossil fuel combustion, cement production, agricultural activities, and natural processes. Emphasis is on the major role that human activities play in causing the rise in global temperatures.	Natural Resources; Natural Hazards; Human Impacts on Earth Systems; Global Climate Change	Cite specific text; integrate information in text with visuals; write arguments; write informative/explanatory texts; conduct short research projects; gather information; draw evidence to support research; reason abstractly and quantitatively; understand ratio language; use variables and write equations.
MS-ETS1 Engineering Design Emphasis on defining design problem to ensure successful solution, *taking into account relevant scientific principles and potential impacts on people and the natural environment that may limit possible solutions;* evaluating solutions; testing among several design solutions and developing model for iterative testing and modification of proposed tool, object, or process *such that optimal design can be achieved.*	Defining and Delimiting Engineering Problems; Developing Possible Solutions; Optimizing the Design Solution	Cite specific text; integrate information in text with visuals; compare and contrast information; conduct short research projects; gather information; draw evidence to support research; reason abstractly and quantitatively; solve problems with positive and negative rational numbers; develop a probability model and use it to find probabilities of events.

SOURCE: NGSS Lead States. (2013). *Next Generation Science Standards: For States, By States.* Washington, DC: The National Academies Press.

Ideally, the above performance expectations can be augmented and enriched through interdisciplinary studies that help frame the societal dynamics driving per capita consumption. At this stage, as students beginning to understand the carbon cycle and how humans fit into it, having students conduct an analysis of their own carbon footprint (usually at the family level) can be an informative exercise that engages students in entering data and then considering the output. (CLEAN has several options in its catalog.) However, a few caveats are worth highlighting:

- Most carbon calculators are black boxes that provide little or no transparency on how the final output is calculated. This is a missed teachable moment as I've explored in this activity on comparing carbon calculators: http://serc.carleton.edu/introgeo/teachingw data/examples/54856.html.

- The final number—usually given in tons of carbon dioxide—should be merely the beginning of further examination of the societal dynamics and economic/ecological issues at play.

- Almost without exception, more affluent students will have a higher footprint than those who are less affluent or poor. This, too, is a potential teachable moment, but one that should clearly be handled with care to avoid calling attention to those students who are not well off or demonizing affluent students, none of whom had a choice in what family they became part of.

HIGH SCHOOL: MASTERING SKILLS, MEASURING RISKS, AND SOLVING PROBLEMS

The Next Generation Science Standards focus on physical, life, and Earth and space sciences, but in many respects they cover the nexus of climate, energy, and water. Assuming they are widely adopted and deployed, by graduation from high school students should have experience in the following:

- Analyzing energy transfer, conversion, and distribution through various systems—including both natural and built systems

- Examining the flow and feedbacks of energy in the climate system, how these processes influence climate over powers-of-ten timescales, and how life has coevolved with the Earth system

- Modeling the cycling of carbon through the Earth system, including human activities

- Evaluating and presenting evidence on how human activities influence climate and how changes in climate have influenced human activity

- Conducting cost benefit analysis for fossil fuel extraction, including tar sands and oil shale, and agricultural practices, emphasizing conservation, recycling, and reuse of resources

- Closely studying solutions to minimize human impacts of natural systems, ranging from local efforts to large-scale geoengineering designs

- Analyzing, designing, and evaluating solutions for a complex real-world problem/global challenge

That would seem to be a tall order, clearly beyond the ability of most current high school students (and many teachers), but most of the elements to facilitate such learning already exist in the CLEAN catalog and elsewhere.

Physical science performance expectations bridge physics and engineering principles, particularly as they require students to develop strategies to conserve energy, crosscutting with relevant concepts and practices.

In life sciences, the performance expectations weave together mathematical reasoning and tools to examine biodiversity and ecosystems, with design solutions to minimize human impacts on the environment.

As for Earth and space sciences, the Earth system is investigated in depth, with the mechanisms and implications of climate change placed front and center. Performance expectations associated with **ESS3: Earth and Human Activities** probe natural resources used by humans, natural hazards that impact them, the ways humans affect the Earth systems, and global climate change. In many ways, the culmination of prior investigations of systems, including engineering and technology practices, analysis and modeling of data, mathematical thinking, and developing solutions to the "many challenges facing long-term human sustainability on Earth" are brought together through the crosscutting concepts and engineering and scientific practices.

| --- | --- | --- |
HS-PS3 Energy Emphasis on calculating flow and change of energy in a system; modeling energy at a macroscopic scale; designing, building, and refining a device converting energy from one form to another; examining thermal energy transfer (second law of thermodynamics) and interactions of two objects through electric or magnetic fields.	Related Disciplinary Core Ideas: Definitions of Energy; Conservation of Energy and Energy Transfer; Relationship Between Energy and Forces; Defining and Delimiting Engineering Problems	Cite specific text; conduct short research projects; gather information; draw evidence to support research; make strategic use of digital media; reason abstractly and quantitatively; model with mathematics; use units; define appropriate quantities; choose appropriate level of accuracy.
HS-LS1 From Molecules to Organisms: Structures and Processes Most relevant to climate and energy is focus on <u>how photosynthesis transforms light energy into storage of chemical energy; how carbon, hydrogen and oxygen form sugar molecules that may combine with other elements;</u> and how cellular respiration as a chemical process results in a net transfer of energy.	Structure and Function; Growth and Development of Organisms; Organization for Matter and Energy Flow in Organisms	Cite specific text; write informative/explanatory texts; develop and strengthen writing; conduct short and more sustained research projects; gather information; draw evidence to support research; make strategic use of digital media; model with mathematics; graph and write functions.
HS-LS2 Ecosystems: Interactions, Energy, and Dynamics Emphasis on factors that affect carrying capacity of ecosystems at different scales; factors affecting biodiversity and populations; cycling of matter and flow of energy in aerobic and anaerobic conditions; <u>the role of photosynthesis and cellular respiration in the cycling of carbon among the biosphere, atmosphere, hydrosphere, and geosphere; how changing conditions may result in new ecosystem, design, evaluation and refinement of a solution for reducing human impacts on the environment and biodiversity;</u> and evaluating the role of group behavior on individual and species' changes to survive and reproduce.	Interdependent Relationships in Ecosystems; Cycles of Matter and Energy Transfer in Ecosystems; Ecosystem Dynamics, Functioning, and Resilience; Social Interactions and Group Behavior; Biodiversity and Humans; Energy in Chemical Processes; Developing Possible Solutions	Cite specific text; integrate and evaluate multiple sources of information; assess reasoning and evidence; evaluate hypothesis, data, analysis, and conclusions of text; write informative/explanatory texts; develop and strengthen writing; conduct short and more sustained research projects; reason abstractly and quantitatively; model with mathematics; use units; define appropriate quantities; choose level of accuracy; represent data; understand statistics, evaluate reports.
HS-ESS2 Earth's System Most relevant to climate and energy is the <u>analysis of geoscience data on feedbacks that cause changes to other Earth systems; flow of energy into and out of Earth's system resulting in changes of climate</u> (using powers-of-ten temporal scaling); examining the properties of water and its effects on Earth materials and surface processes; the cycling of carbon among the hydrosphere, atmosphere, geosphere, and biosphere; and the coevolution of Earth's systems and life on Earth (such as how *photosynthetic life altered the atmosphere through the production of oxygen, which in turn increased weathering rates and allowed for the evolution of animal life*).	Earth and the Solar System; Earth Materials and Systems; Plate Tectonics and Large-Scale System Interactions; The Roles of Water in Earth's Surface Processes; Weather and Climate; Biogeology; Wave Properties	Cite specific text; determine central idea of text; write arguments focused on discipline-specific content; conduct short and more sustained research projects; make strategic use of digital media; reason abstractly and quantitatively; model with mathematics; use units; define appropriate quantities; choose level of accuracy.

(Continued)

PERFORMANCE EXPECTATION	RELATED DCI	LINKS TO COMMON CORE
HS-ESS3 Earth and Human Activity Emphasis on _how the availability of natural resources, occurrence of natural hazards and changes in climate have influenced human activity_; evaluating design solutions relating to the extraction and consumption of energy and mineral resources; the relationships between natural resources, sustainability of human populations and biodiversity, technological solutions to reduce human activities on natural systems (local to global scale); _analyzing geoscience data and global climate models to make evidence based forecast of the current rate of global or regional climate change and associated future impacts to Earth systems_; and illustrating the relationships _among Earth systems and how those relationships are being modified due to human activity._	Weather and Climate; Natural Resources; Natural Hazards; Human Impacts on Earth Systems, Global Climate Change; Developing Possible Solutions	Cite specific text; determine central idea of text; integrate and evaluate multiple sources of information; evaluate hypotheses, data, analysis, and conclusions; write informative/explanatory texts; reason abstractly and quantitatively; model with mathematics; use units; define appropriate quantities; choose level of accuracy.
HS-ETS1 Engineering Design Emphasis on analysis of criteria and constraints of solutions for a major global challenge; _design solution to a complex real-world problem by breaking it down into smaller, more manageable problems that can be solved through engineering; and evaluating cost, safety, reliability, aesthetics and potential social, cultural and environmental impacts of a solution, i.e., risks._	Defining and Delimiting Engineering Problems; Developing Possible Solutions; Optimizing the Design Solution	Links to Common Core: Integrate and evaluate multiple sources of information; evaluate hypotheses, data, analysis, and conclusions; synthesize information from a range of sources; reason abstractly and quantitatively; model with mathematics.

SOURCE: NGSS Lead States. (2013). _Next Generation Science Standards: For States, By States_. Washington, DC: The National Academies Press.

If successfully implemented, NGSS will help young people better understand and be able to act in an informed way to the causes, effects, risks, and potential responses to human-induced climate change. These young people will ultimately be better prepared to address future challenges in an informed way, thus helping the nation and ultimately the people of the planet. Beyond mastering the crosscutting concepts and developing the necessary scientific, technical, and collaborative skills, learners—and by extension their families, educators, and their supporting infrastructure and society at large—will need to make addressing human impacts to global change a primary and compelling priority of the 21st century.

Is NGSS Enough?

The framework states that "by the end of the 12th grade, students should have gained sufficient knowledge of the practices, crosscutting concepts, and core ideas of science and engineering to engage in public discussions on science-related issues, to be critical consumers of scientific information related to their everyday lives, and to continue to learn about science throughout their lives. They should come to appreciate that science and the current scientific understanding of the world are the result of many hundreds of years of creative human endeavor. It is especially important to note that the above goals are for all students, not just those who pursue careers in science, engineering, or technology or those who continue on to higher education." But is this true for climate and energy science?

Are the NGSS enough to fully prepare learners to understand climate and energy issues? The Thomas B. Fordham Institute, which gave NGSS a "C" rating, claiming that many states have better existing standards (but ignoring that existing standards don't necessarily reach all students), has been critical of the way climate change has been handled in the NGSS (Gross et al., 2013).

> While acknowledging that NGSS's inclusion of expectations that address climate change is both useful and touchy, the Fordham critique suggests that to deal competently with the content at the high school level, students need a deeper understanding of chemistry (including pH of the ocean, isotopic dating of ice cores) and physics (including the mechanism of the greenhouse effect) as well as the use of computer models. Noting that high school students *could* develop "an elementary but realistic sense of "climate science" if coherently developed over time, they conclude that "little such development is visible in these standards."

In fairness, the NGSS represents an effort to raise the floor for all students, but it does not impose a ceiling. Incorporating the concepts and principles from the climate literacy and energy literacy frameworks is a one way to raise the ceiling, as we'll see in the Chapters 4 and 5, which feature numerous activities for educators to use in the their classrooms, providing practical and interactive ways to teach climate and energy science.

ADDITIONAL RESOURCES

NGSS. (2014). *Three dimensions of next generation science standards.* Retrieved from http://www.nextgen science.org/three-dimensions
> *This is a two-page summary of the practices, DCI, and crosscutting themes of NGSS.*

Willard, T. (2013). *Inside the NGSS.* Retrieved from http://www.aft.org/pdfs/teach2013/ NextGenerationScienceStandardsNGSSBox_MH.pdf
> *This is a two-page, inside-the-box summary examining the architecture of the Next Generation Science Standards from an expert who helps examine the NGSS ecosystem.*

CHAPTER 4

Teaching Climate Literacy

How is climate science literacy defined? Very simply, it is "an understanding of the climate's influence on you and society and your influence on climate" (NOAA, 2009). Ideally, by understanding the essential principles that make up climate science, knowing how to assess credible scientific information about climate, and communicating about it in a meaningful way, a climate-literate person will also "be able to make informed and responsible decisions with regard to actions that may affect climate" (NOAA, 2009).

Since 2009, many climate change education projects and programs have directly or indirectly used the *Climate Literacy* framework to help identify areas to focus on, including CLEAN, the Climate Literacy & Energy Awareness Network. CLEAN has done considerable work to review thousands of online resources, catalog and annotate many of the very best, and tag them not only to the *Climate Literacy* principles but also to the *Energy Literacy* principles we'll examine in Chapter 5. CLEAN serves as a clearinghouse and "referatory" for educational resources vetted for accuracy and pedagogical quality with the purpose of building understanding of the core ideas in climate and energy science.

BEGINNINGS

In a St. Petersburg, Florida, hotel conference room in 2005 as part of a Digital Library for Earth Science Education (DLESE) annual meeting while Hurricane Dennis roared offshore toward the Florida panhandle, Frank Niepold and I met for the first time. A former classroom science teacher and father of two young sons, Niepold was working at NASA and teaching high school Earth system science at the time but heading to work for the science and education community at NOAA as the first NOAA Climate Education Fellow. We shared an interest in identifying the very best online resources about climate but also in providing teachers and learners with the "essential principles and fundamental concepts" relating to climate in general and human impacts on the climate system in particular. We discussed the need for a framework on climate science similar to the Ocean Literacy framework that NOAA, the National Geographic Society, and the National Marine Educators Association had developed.

Niepold eventually became the communication and education cochair at U.S. Global Change Research Program, and the NOAA climate education coordinator, and working together with other dedicated educators and scientists, we released the first version of *Climate Literacy* in 2008 for K-12 Teachers. We also formed the Climate Literacy Network to help support federal climate education efforts and coordinate efforts outside of the government.

Revisiting the first draft and broadening it for "individuals and communities" rather than focusing only on teachers, we were able, thanks largely to Frank's efforts, to get the current version approved by the U.S. Global Change Research Program, and in March 2009 at the National Science Teacher Association (NSTA) national conference, the current document was released. It is available online through the USGCRP website at http://globalchange.gov/resources/educators, and excerpts of the text are included in Appendix II.

Tamara Ledley of the Cambridge, MA-based TERC (Technical Educational Research Centers) became the chair of the Climate Literacy Network, which eventually was rebranded as CLEAN, and weekly teleconference calls were held every Tuesday beginning in early 2008 to develop and support the *Climate Literacy* framework. In early 2009, Congress authorized funding for climate change education grants, primarily through the NSF and NASA, and for a time, the drought of federal funding to develop resources and professional development about climate science was over. Tamara and I, along with Frank and several others—with NSF support—developed the CLEAN collection of reviewed and annotated online resources that tie to climate or energy literacy frameworks, and many of these resources are highlighted in this and the following chapter.

In 2013 we established the National Climate Assessment Network (NCAnet) Education Affiliate Group, which has developed "treasure maps" to help educators and learners discover and dig out the treasures of the National Climate Assessment (NCA), released in May, 2014. A veritable treasure trove of information, images, and data designed to be accessible for mobile learning through smart phones and tablets, the NCA has the potential to help transform climate education in the United States. NCA is available at http://nca2014.globalchange.gov/ and the guides for educators are available at http://www.climate.gov/teaching.

CLEAN Resources, the NRC Framework for K–12 Science Education, and the NGSS

On its website, CLEAN includes maps of climate and energy concepts that use the National Science Digital Library Strand Maps to show where concepts fit into learning development and to point to relevant resources. It also offers a detailed section on teaching about climate, which is designed specifically to guide teachers through what the climate essential principles mean, why they are important, what makes particular principles challenging to teach, and how the principles can be woven into teaching middle and high school as well as college undergraduate students. Relevant online resources in CLEAN for these educational levels are included as well as references for further exploration. Rather than recreate the excellent content in CLEAN, in this chapter the key ideas relevant to teaching and learning each of the climate principles will be summarized with links cited to the CLEAN website where more information can be accessed.

The intent and spirit of NGSS and the NRC K–12 Framework are to foster not only content understanding but also the skills and knowledge needed to make informed decisions. Although human impacts to the climate system and responses to minimize and prepare for the resulting changes that are already occurring are not officially introduced until middle school in the NGSS, the groundwork for considering how humans, like other organisms, can alter our surroundings and what can be done to reduce the negative impacts is introduced in a properly gentle way in kindergarten.

Following is a table that provides an initial overview of relationships among the *Climate Literacy* framework's principles, climate and energy concepts (presented here in an abbreviated form), disciplinary core ideas identified in the Framework for K–12 Science Education, and performance expectations described in the NGSS. Corresponding CLEAN web resource links are also included that demonstrate how these ideas and expectations are interwoven. In the table, summaries of the concepts are shortened from the original literacy document. Excerpts from the *Climate Literacy* guide are available in Appendix II.

Before we review the individual essential principles and summarize their respective concepts, a few words about their organization, which is in a hierarchical, linear manner, starting with the first principle, the Sun is the primary source of energy for Earth's climate system; moving into the workings of the Earth system, how scientists know what they know; and culminating in the final principle, climate change will have consequences for the Earth system and human lives. NGSS, which starts simply and then builds on prior knowledge, has a somewhat similar general flow, starting simply with observations of the role of the Sun and seasons in our lives and in ecosystems, gradually adding more levels of complexity. By the time learners achieve high school graduation and are moving on to college or careers, they should have a solid understanding of the Earth system and its influence on humanity as well as how human impacts are altering the planet and what can be done to minimize and respond to the negative effects.

There will be parts of some of the principles that can be explored in elementary grades, especially as they relate to weather and climate, such as the fourth principle, which covers the concept that climate, while related, is not the same thing as weather; weather occurs over hours and days, climate over seasons and years. Also appropriate in primary grades is a general exploration of ecosystems and how they depend on, are shaped by, and to some degree, affected by climate, which is the focus of Principle 3.

ESSENTIAL PRINCIPLES	VITAL CONCEPTS (IN BRIEF)	DISCIPLINARY CORE IDEAS (PRIMARY) FROM THE NRC K-12 FRAMEWORK FOR SCIENCE EDUCATION	NGSS PERFORMANCE EXPECTATIONS (OVERVIEW)
Climate Literacy (with links to CLEAN-related teaching resources)	Essential Principles of Climate Science	Weather & Climate Global Climate Change Human Impacts on Earth Systems	A significant number of performance expectations relate to climate literacy, beginning in kindergarten.

ESSENTIAL PRINCIPLES	VITAL CONCEPTS (IN BRIEF)	DISCIPLINARY CORE IDEAS (PRIMARY) FROM THE NRC K-12 FRAMEWORK FOR SCIENCE EDUCATION	NGSS PERFORMANCE EXPECTATIONS (OVERVIEW)
Guiding principle for informed climate decisions: Humans can take actions to reduce climate change and its impacts. CLEAN Link: http://cleanet.org/clean/literacy/guiding_principle.html	Information and understanding can reduce vulnerability or enhance resilience, improve decision-making; mitigation through reduced impacts and adaptation are both required; decisions occur at all levels of society, impacting opportunities and limits of current and future generations.	Physical Science Life Science Earth & Space Science Engineering, Technology & Applications of Science	Particular emphasis on 3-5 ETS* engineering design problems and solutions and influence on society and natural world.
1. The Sun is the primary source of energy for Earth's Climate System. CLEAN Link: http://cleanet.org/clean/literacy/principle_1.html	Earth's energy budget/balance; tilt of axis as reason for the seasons; orbital changes (long-time scales); short-term solar variability doesn't account for recent warming.	EES1.A: The Universe & Its Stars ESS1.B: Earth & the Solar System	Begins with K-PS3-1 (observations of sunlight) through HS-ESS3-6 (show how human activity are modifying Earth systems).
2. Climate is regulated by complex interactions among components of the Earth system. CLEAN Link: http://cleanet.org/clean/literacy/principle_2.html	Interactions within the system and local climate variations; ocean role in climate system; greenhouse effect (basic dynamics and more complex chemistry/physics); aerosols/dust; feedbacks.	ESS1.B: Earth & the Solar System ESS2.A: Earth Materials & Systems ESS2.C: The Roles of Water in Earth's Surface Processes ESS2.D: Weather & Climate	From 5-ESS2-1 (ways Earth systems interact) to MS-ESS2-4 (complex interactions of air masses in changing weather conditions) and HS-ESS2-2 (climate feedbacks).
3. Life on Earth depends on, is shaped by, and affects climate. CLEAN Link: http://cleanet.org/clean/literacy/principle_3.html	Natural greenhouse effect allows liquid water and life to exist; life forms are integral to carbon cycle and climate system; organisms and ecosystems adapt to, migrate from, or perish when climate changes; abrupt extinctions have occurred in the past; human societies have developed and flourished since end of last Ice Age.	LS2A: Interdependent Relationships in Ecosystems LS2.C: Ecosystem Dynamics, Functioning, & Resilience LS4.C: Adaptation LS4.D: Biodiversity & Humans ESS2.E: Biogeology	Begins with K-ESS2-2 (how plants and animals can change environment) to MS-LS2-5 (design solutions for biodiversity and ecosystem services) and MS-LS1-6 (role of photosynthesis in cycling of matter and flow of energy), to HS-LS2.5 (role of photosynthesis in carbon cycling), HS-ESS2-6 (model carbon cycling), and HS-ESS2-7 (coevolution of Earth's systems and life).
4. Climate varies over space and time through both natural and man-made processes. CLEAN Link: http://cleanet.org/clean/literacy/principle_3.html	Difference and relationship between weather and climate; climate change is more than normal variations (seasons, ENSO); climate change does occur naturally, but recent changes can't be explained without including human impacts, especially CO_2 from burning fossil fuels.	ESS1.C: The History of Planet Earth ESS2.C: The Roles of Water in Earth's Surface Processes ESS2.D: Weather & Climate ESS3.C: Human Impacts on Earth System ESS3.D: Global Climate Change	From K-ESS2-1 (weather patterns over time) and 2-ESS-1 (events can occur quickly or slowly on Earth), to 5-ESS1-2 (patterns of seasonal changes), MS-ESS2-2 (geoscience processes over time and spatial scales), and HS-ESS2-4 (climate changes over powers-of-ten tie scales).

(Continued)

ESSENTIAL PRINCIPLES	VITAL CONCEPTS (IN BRIEF)	DISCIPLINARY CORE IDEAS (PRIMARY) FROM THE NRC K-12 FRAMEWORK FOR SCIENCE EDUCATION	NGSS PERFORMANCE EXPECTATIONS (OVERVIEW)
5. Our understanding of the climate system is improved through observations, theoretical studies, and modeling. CLEAN Link: http://cleanet.org/clean/literacy/principle_5.html	How scientists measure, analyze and project future climate; weather models are very different from climate models; current projections are sufficient to evaluate decision-options and climate-change responses.	ESS1.C: The History of Planet Earth ESS2.D: Weather & Climate ESS3.C: Human Impacts on Earth System ESS3.D: Global Climate Change ETS2.B: Influence of Science Also, strong tie to Scientific & Engineering Practices	Beginning with observations of local weather in kindergarten, scientific & engineering practices are emphasized throughout the grade bands, culminating in HS-ETS1 Engineering Design, where the focus is on addressing complex real-world problem.
6. Human activities are impacting the climate system. CLEAN Link: http://cleanet.org/clean/literacy/principle_6.html	Strong agreement in climate research community that human activities, especially burning fossil fuels and resulting release of CO_2, are altering the planet; other human impacts include releasing chemicals, changing land cover, impacting habitat; warming from industrial development will last for centuries; likely more negative than positive impacts if warming exceeds 2-3C (3.6-5.4F).	ESS3.C: Human Impacts on Earth System ESS3.D: Global Climate Change	Beginning with K-ESS3-3 (reducing impact of humans on local environment), learners focus more specifically on human impacts in middle school (MS-ESS3-5, factors that have caused the rise in global temperatures) and into high school (HS-ESS3-5, analyzing forecasts of climate change).
7. Climate change will have consequences for the Earth system and human lives. CLEAN Link: http://cleanet.org/clean/literacy/principle_7.html	Impacts include sea level rise, changes in fresh water resources, ocean acidification, heat waves, dry regions becoming dryer, wetter regions becoming wetter, altered ecosystems, human health and mortality rates affected, particularly for most vulnerable populations.	ESS3.C: Human Impacts on Earth System ESS3.D: Global Climate Change	The ways human activities other than climate change can influence ecosystems is established in elementary grades. In middle school, students examine how population and consumption of natural resources impact Earth systems (MS-ESS3-4), and in High School they look into how climate change and other factors influence human activities (HS-ESS3-1).

NOTE: * ETS refers to the Engineering, Technology, and Applications of Science

FOCUS ON THE "GUIDING PRINCIPLE"

One of the most forward-leaning aspects of the *Climate Literacy* framework when it was released was the "Guiding Principle for Informed Climate Decisions." The principle itself is simple and straightforward: Humans can take actions to reduce climate change and its impacts.

Often overlooked, the concepts that buttress the "Guiding Principle" are clear, self-evident, and to the point. They are included below in full to emphasize their importance:

A. Climate information can be used to reduce vulnerabilities or enhance the resilience of communities and ecosystems affected by climate change. Continuing to improve scientific understanding of the climate system and the quality of reports to policy and decision makers is crucial.

B. Reducing human vulnerability to the impacts of climate change depends not only upon our ability to understand climate science but also upon our ability to integrate that knowledge into human society. Decisions that involve Earth's climate must be made with an understanding of the complex interconnections among the physical and biological components of the Earth system as well as the consequences of such decisions on social, economic, and cultural systems.

C. The impacts of climate change may affect the security of nations. Reduced availability of water, food, and land can lead to competition and conflict among humans, potentially resulting in large groups of climate refugees.

D. Humans may be able to mitigate climate change or lessen its severity by reducing greenhouse gas concentrations through processes that move carbon out of the atmosphere or reduce greenhouse gas emissions.

E. A combination of strategies is needed to reduce greenhouse gas emissions. The most immediate strategy is conservation of oil, gas, and coal, which we rely on as fuels for most of our transportation, heating, cooling, agriculture, and electricity. Short-term strategies involve switching from carbon-intensive to renewable energy sources, which also requires building new infrastructure for alternative energy sources. Long-term strategies involve innovative research and a fundamental change in the way humans use energy.

F. Humans can adapt to climate change by reducing their vulnerability to its impacts. Actions such as moving to higher ground to avoid rising sea levels, planting new crops that will thrive under new climate conditions, or using new building technologies represent adaptation strategies. Adaptation often requires financial investment in new or enhanced research, technology, and infrastructure.

G. Actions taken by individuals, communities, states, and countries all influence climate. Practices and policies followed in homes, schools, businesses, and governments can affect climate. Climate-related decisions made by one generation can provide opportunities as well as limit the range of possibilities open to the next generation. Steps toward reducing the impact of climate change may influence the present generation by providing other benefits, such as improved public health infrastructure and sustainable built environments.

As Mary Pipher writes in *The Green Boat: Reviving Ourselves in Our Capsized Culture* (2013):

Climate educators must balance information with action suggestions, motivational elements, and aspirational framing. If we want people to listen to and process traumatic information, then we must be able to frame that information in ways that allows listeners to be hopeful and calm. (p. 26)

Both the *Climate Literacy* framework and the NGSS provide this pragmatic mix of science and solutions, risks and responses without political or ideological agenda.

Tempting though it may be, especially for those who work with younger learners, to offer simple solutions to avoid too much gloom and doom, the causes, effects, risks, and responses to climate change are inherently complex. In addition, many of the issues touched on in the concepts above go beyond the purview of the science classroom and should ideally be taught either through team teaching or in social studies, civics, geography, or other classes, otherwise they may distract from the main focus of a science class: mastering the science. But education is as much an art as a science, so educators should trust their experience and intuition to find the most effective and appropriate way to engage their learners.

Whether it's with the "Guiding Principle for Informed Climate Decisions" or the "Essential Principles" we'll be examining next, teaching and learning are often challenged by what are variously described as misconceptions, naive concepts, preconceptions, and what some may call misinformation. For the sake of simplicity, these have been lumped together under the heading of Missed Conceptions. Whatever their cause, research and experience make it clear that they can be difficult to replace with the "right" concepts for a range of reasons. These may include (but are not limited to) that the "right" concept may be complex and nonintuitive, the "wrong" answer is stickier in the mind of the learner than the correct one, the learner may not have sufficient background experience to be able to grasp the correct knowledge, and it is also possible to have multiple concepts in the mind—some accurate, some not—at the same time. For example, it is common knowledge that the Sun doesn't revolve around the Earth and that it takes the Earth a year to circle the Sun, yet in our everyday experience and even language it seems as though the Sun rises in the East and sets in the West, as if the Earth were standing still.

And just a quick format note for the remainder of this chapter—after elaborating on each principle and then following that elaboration with the missed conceptions, we'll take a look at ways of measuring or mathematically calculating information related to the principle and related concepts in a section called Doing the Math. That will be followed by a brief list, by grade level, of CLEAN-recommended interactive resources and activities to enrich and extend the learning.

DOING THE MATH

Measuring the causes, effects, risks, and responses to climate change requires measuring the physical processes and trends of the climate system and human contributions as well as the more intangible factors of human behavior, which can be mathematically mapped to a large degree but include wildcard uncertainties. Among the unknowns is the speed that climate and energy literacy can be ramped up in society and the extent that improved literacy will help in reducing impacts and vulnerabilities and accelerating innovation and social entrepreneurship.

CLEAN Recommends

Middle School

> Clarkson Energy Choices Board Game—Clarkson and St. Lawrence Universities
>> http://cleanet.org/resources/41837.html

> The Big Energy Gamble—Jeff Lockwood, NOVA Teacher
>> http://cleanet.org/resources/41864.html

High School

> Stabilization Wedges Game—Carbon Mitigation Initiative, Princeton University
>> http://cleanet.org/resources/41709.html

Introductory College

> The Lifestyle Project—Karin Kirk, John J. Thomas, SERC—Starting Point Collection
>> http://cleanet.org/resources/41895.html

FOCUS ON THE SEVEN ESSENTIAL PRINCIPLES

The "Essential Principles" are the underpinning of the *Climate Literacy* "Guiding Principle" that humans can take actions to reduce climate change and its impacts. Each principle adds a dimension to our understanding of climate and energy science.

1. THE SUN IS THE PRIMARY SOURCE OF ENERGY FOR EARTH'S CLIMATE SYSTEM.

There are five key concepts that nest within the first principle that can be summarized in the following way:

A. Incoming sunlight reaches Earth.

B. There is a relationship between incoming and outgoing energy that relates to its energy balance and budget.

C. Factors—specifically the tilt of the Earth on its axis—alter the amount of sunlight received on different parts of the planet, thereby changing temperatures.

D. Gradual changes in the planet's orbit can drive long-term changes in climate.

E. Changes in the Sun's energy output can cause the Earth to warm or cool, but recent warming is not related to such solar variability according to detailed analysis of such factors.

In elementary grades, the focus on sunlight and learning about the basics of energy is the appropriate focus. But in science classes many of the nuances of these concepts are advanced and often not covered or are skimmed over. For example, the long-term changes in the planet's rotation and orbit described in Concept D, which are known as

Milankovitch Cycles in paleoclimatology, are often not taught and may be challenging for even undergraduate students to master. The "axial tilt as the reason for the seasons" covered in Concept C may be taught in middle school, but developmentally it is more appropriate for high school, where it is rarely revisited. And the details of the Earth's energy budget touched on in Concept B require an understanding of the greenhouse effect, which is mentioned but not fully described in Essential Principle 2, C and D.

There are ample opportunities to weave solutions and responses to climate and global change into teaching about this first principle, such as discussions about the use of passive and active solar design in architecture, building solar hot water heaters, or examining photovoltaic and concentrated solar power generation of electricity in secondary schools, and even research on quantum photosynthesis (Mundet, 2013) in college.

Missed Conceptions

- Wavelength characterization of energy from the Sun (mostly visible, with some ultraviolet and shortwave infrared)

- Role and nature of atmosphere in filtering incoming and outgoing energy—air has mass and density, even though we generally don't notice in our everyday experience, just as fish in the ocean pay no attention to the pressure of water

- Role of albedo (reflectivity) in the Earth's energy budget, confusing this with absorption of energy

- The axial tilt as the reason for the seasons (not closeness to the Sun)

- The Earth's orbit around the Sun (only slightly elliptical, despite visualizations to the contrary) and long-term changes to the wobble of Earth on its axis, which over time can vary between 22.1° and 24.5°, currently being 23.5°, and small but important changes in the eccentricity—deviation from a perfect circle—of the Earth's orbit of the Sun

- The mechanics and role of the greenhouse effect, particularly the fact that the Earth receives nearly twice as much heat from the greenhouse effect as it does from the incoming energy from the Sun

- Confusing the ozone layer with global warming

Because this last point is common, it is worth taking a moment to unpack it. Many people confuse the human threat to the ozone layer—a major concern in the 1980s leading to the Montreal Protocol, which phased out ozone-depleting chemicals—to the human impact on climate. Some textbooks even combine discussion of the ozone layer, which has been depleted because of human-generated gases, such as chlorofluorocarbons (CFCs), with climate change, which is primarily caused by carbon emissions, in the same chapter, blurring the differences between the two issues. The mental model stems from the idea that a hole in the atmosphere will let more heat from the Sun to reach the Earth, thereby warming it. The confusion is further complicated by the fact that some CFC-like gases are also greenhouse gases and that ozone (O_3) in the troposphere can act as a greenhouse gas. On top of that, the ozone holes, which are more accurately described as thinning of the layer primarily in the polar regions, appear to affect the polar vortices (Holdren, 2014), which are large-scale cyclones that alter wind patterns.

An interesting note: Some progress has been made in protecting the ozone layer through controlling chemicals that deplete the ozone layer because of the passage of the Montreal Protocol, the first treaty in the history of the United Nations that has been universally adopted. Many had hoped that the Kyoto Protocol, modeled in part on the Montreal Protocol, would be a similar international treaty to limit greenhouse gas emissions, but as philosopher Dale Jamieson details in his book *Reason in a Dark Time: Why the Struggle Against Climate Change Failed—and What It Means for Our Future* (2014), the challenges of regulating fossil fuel emissions in particular at an international level have proven much more daunting than regulating ozone-depleting chemicals.

Doing the Math

There is no shortage of ways to measure these concepts. The NGSS begin with simple measurements of sunlight and weather in the early primary grades and move into increasingly complex observations and data collection/analysis in the secondary grades. These geophysical measurements can include tracking seasonal changes in the length of days and related climate and environmental factors, such as changes in vegetation or migration patterns and measuring precipitation, temperature, and air pressure, which helps in emphasizing the fact that air, which seems weightless in our everyday experience, has mass and is subject to the laws of physics. For the math-inclined student, delving into orbital factors, such as axial tilt, eccentricity, and the corresponding changes in solar input, lend themselves to global climate modeling and related analysis.

CLEAN Recommends

Middle School

My Angle on Cooling: Effects of Distance and Inclination—AAAS Science NetLinks

http://cleanet.org/resources/41806.html

Earth's Energy Cycle—Albedo, National Center for Atmospheric Research (NCAR)

http://cleanet.org/resources/41824.html

High School

Effect of the Sun's Energy on the Ocean and Atmosphere—Mitch Fox, NASA Goddard Space Flight Center

http://cleanet.org/resources/41892.html

Introductory College

The Earth's Heat Budget—Roy Plotnick, National Association of Geoscience Teachers

http://cleanet.org/resources/41888.html

Early climate investigators such as Fourier and Tyndall correctly deduced that there was an atmospheric component of the climate system that kept the planet warmer than it would be otherwise, and Tyndall found two of the key factors that absorb infrared energy: carbon dioxide and water vapor. Here is a walk-through of the famous but rarely understood or sufficiently described greenhouse effect:

1. Sun emits energy. The Sun is a sphere of gases and plasma that radiates intense energy in the electromagnetic spectrum. Most of the energy reaching Earth is in and around wavelengths in the visible spectrum (between ultraviolet and shortwave infrared), and some of the incoming energy is absorbed or scattered by the Earth's atmosphere before reaching the surface.

2. Energy reaches Earth's surface. Energy reaching the Earth's surface drives photosynthesis, the hydrologic cycle, and other physical processes. Earth's temperature increases when the energy input exceeds the energy output.

3. Earth's surface absorbs energy. A variety of substances cover Earth's surface—including land, water, ice, and biologic organisms. These reflect or absorb and transform varying amounts of the Sun's energy. Organisms that absorb sunlight through photosynthesis form the base of the food chain.

4. Earth's surface radiates the energy as heat. Earth's surface absorbs roughly half the incoming solar radiation, increasing its temperature. Warmed by the Sun, the Earth's surface radiates the energy that it has absorbed in the form of longwave infrared energy or heat. This dynamic can be experienced when rocks or pavement exposed to the Sun during the day release heat at night.

5. Heat is absorbed by greenhouse gasses in the atmosphere. Ninety percent of the infrared light radiating from the Earth's surface is absorbed by specific atmospheric gases—the so-called *greenhouse gases*. These gases are essentially transparent or invisible to the shortwave incoming energy from the Sun but absorb the outgoing longwave infrared energy then radiate it back at infrared wavelengths. Some of the radiation from the greenhouse gases radiates back down toward the surface, serving as an additional heat input.

6. The greenhouse effect is natural. This phenomenon is called the *greenhouse* effect, even though it is a different process than an actual greenhouse, in which heat builds up because it is trapped inside an enclosure. Because of this process, the surface of the planet is about 33 °C (60 °F) hotter than it would be without greenhouse gases in the atmosphere, and this allows liquid water and "life as we know it" to exist.

7. Not all greenhouse gases are equal. Water vapor, which is a powerful greenhouse gas, cycles through the atmosphere in a matter of days or weeks, while other greenhouse gases like carbon dioxide or methane may remain in the atmosphere for years and even centuries, because the natural processes that cycle them are much slower than the precipitation-evaporation-condensation processes of the water cycle. Carbon dioxide, the most important of the long-lasting gases, is used as a benchmark to compare the global warming potential (GWP) of the other gases. Methane is another important greenhouse gas that is emitted naturally from wetlands and through human activities, such as livestock and agricultural production. Human-made greenhouse gases, such as certain CFCs, may last thousands of years in the atmosphere.

8. Earth's temperature increases. Human activities, such as burning fossil fuels that release carbon dioxide, add to the concentration of these heat-absorbing gases that absorb the outgoing infrared heat. This generates more heat in the atmosphere, causing Earth's average surface temperature, including the ocean and land surface, to rise, altering ecosystems as a result.

9. It may increase quickly. Current trends indicate that the *planet's global average temperature* (Betts et al., 2011) could rise over 7 °F or 4 °C by the end of this century. It will continue to rise for centuries to come because of the long lifetime of greenhouse gases, although over tens of thousands of years Earth's orbital cycles may eventually counter to some extent the human-induced warming currently being produced.

There are of course detailed nuances to all of the above that are ripe for deeper inquiry: how convection transports energy away from the surface can alter the greenhouse warming; feedbacks from changes in albedo, such as reflective sea ice melting into heat-absorbing

open ocean; and the role of the infrared "water vapor window" in the upper atmosphere to release infrared light to space.

But the basic take-home remains: Higher concentrations of greenhouse gases in the atmosphere, whatever the source, lead to heating of the atmosphere and thereby the planet.

Professor Michael Ranney and colleagues at the University of California at Berkeley have found through research that the understanding of the greenhouse effect is vital to understanding how the climate system works and why humans are altering it (Clark, Ranney, & Felipe, 2013). They have put together a website with a series of videos that range from 52 seconds long to less than five minutes that help explain the basics: http://www.howglobalwarmingworks.org.

An excellent introductory animation about the greenhouse effect is also available through the CLEAN Collection: http://cleanet.org/resources/42851.html.

2. CLIMATE IS REGULATED BY COMPLEX INTERACTIONS AMONG COMPONENTS OF THE EARTH SYSTEM.

The second *Climate Literacy* Principle includes six concepts that point to the following:

A. The ways the Sun, ocean, atmosphere, land, and life interact

B. The role of ocean and hydrologic cycles

C. How small amounts of gases like carbon dioxide can have a large effect on the climate system

D. The role of the carbon cycle in the greenhouse effect

E. The role of particulates (aerosols) on the Earth's energy balance

F. Feedbacks within the various systems that can trigger abrupt, non-linear changes

Key to understanding how the climate system, in all its complexity, works is understanding the fundamentals of the atmosphere and its layers and makeup, the ocean and water cycle, and some basic geography about how land, water, and air relate to each other. Feedbacks, at least the basics of how small forces or factors can cause major impacts, are also vital and can be introduced early on with simple examples, like the animation of a rolling marble from the NOAA Paleo Perspective on Abrupt Climate Change: http://www.ncdc.noaa.gov/paleo/abrupt/story1.html.

While the basics of the hydrologic cycle are generally well covered in science education, the composition and layers of the atmosphere may not be, leading to considerable confusion and misconceptions. After all, air appears invisible and weightless, which may make it difficult to comprehend basic weather and climate processes.

Missed Conceptions

- How seemingly small factors (like concentrations of carbon dioxide in the atmosphere or feedback loops) can trigger major changes in systems

- The vital role of the ocean in storing and releasing heat as well as the long-term carbon cycle

- Role of aerosols (which have nothing to do with hairspray!) from volcanic and human activity in climate dynamics

- The complexity and interconnectivity of systems

Doing the Math

Measuring numerically the flow of energy and matter through various cycles and systems is inherently challenging, but modeling of the Earth system, first qualitatively, conceptually, then in more depth with quantitative rigor will help build analytic and systems-thinking skills that can be transferred to other domains. The Modeling Earth's Climate activity from Concord.org, which is part of the CLEAN collection, can help learners master these systems dynamics.

CLEAN Recommends

As the editors at CLEAN write: "Just because a process is complex does not mean it needs to be complicated" (Clean, n.d.). Here are some of the activities they have called out to help learners master related concepts. Many of the introductory activities could be used in elementary or even as resources for college students whose science background is lacking.

Middle School

Energy Flows—NEED Project—Putting Energy into Education
http://cleanet.org/resources/41911.html

Understanding Albedo—Geophysical Institute at University of Fairbanks
http://cleanet.org/resources/41833.html

High School

Ocean Currents and Sea Surface Temperature—Joan Carter, NASA—*My NASA Data Collection*

http://cleanet.org/resources/41846.html

Introductory College

Tropical Atlantic Aerosols—Rex Roettger, NASA—*My NASA Data Collection*
http://cleanet.org/resources/41842.html

Upper-Level College

Exploring the Link Between Hurricanes and Climate Using GCM Results—Cindy Shellito, SERC—*On the Cutting Edge Collection*
http://cleanet.org/resources/41853.html

3. LIFE ON EARTH DEPENDS ON, IS SHAPED BY, AND AFFECTS CLIMATE.

The third principle brings the life sciences into climate literacy in a major way, with five concepts building on the idea that biologic systems are at once is dependent on, shaped by, and also influence climate. The concepts focus on the following:

A.   How organisms survive or perish, depending on climatic conditions and their ability to adapt or migrate

B.   How liquid water on the planet, necessary for many forms of life to thrive, exists because of the greenhouse effect

C.   How we can see from fossil records how changes in climate can lead to extinctions

D.   How the rise of human civilization has occurred during the past 10,000 years when climate has been relatively stable

E.   That organisms, from microbes billions of years ago to humans today, have influenced climate

Beginning in kindergarten with observations of weather and organisms and building through the grades into more complex observations, analysis, and modeling of changes and processes of climate and ecosystems over time, this principle helps educators and ultimately learners connect the dots between life and physical and life sciences.

Missed Conceptions

- That biology and life sciences are an integral part of climate literacy ("I'd like to teach about climate change, but I'm a biology teacher!")

- That organisms both influence and are influenced by climate

- The role of the carbon cycle as the backbone of life and a driving force in climatic and geologic processes and the timescales involved

Doing the Math

Focused on the biologic dimension to climate, this principle requires, as do all sciences, mathematical calculations to track dynamics, time, and spatial scales, statistics, and probabilities. And as with other sciences, beginning simply and moving into increasingly detailed algebraic dynamics will offer opportunities to link math with bio-climatic interactions.

CLEAN Recommends

Middle School

Blooming Thermometers—Lisa Gardiner et al., National Center for Atmospheric Research (NCAR)

http://cleanet.org/resources/41839.html

Climate Change and Arctic Ecosystems—Project Activities for Conceptualizing Climate and Climate Change, Purdue University

http://cleanet.org/resources/41898.html

High School

Global patterns in Green-up and Green-down—GLOBE Program

http://cleanet.org/resources/41847.html

Introductory College

Understanding the Carbon Cycle: A Jigsaw Approach—David Hastings, SERC—On the Cutting Edge Collection

http://cleanet.org/resources/41877.html

4. CLIMATE VARIES OVER SPACE AND TIME THROUGH BOTH NATURAL AND HUMAN PROCESSES.

There are seven concepts that fall under the fourth principle:

A. How climate is made up of weather patterns over time

B. That weather occurs on a local level over short timescales

C. That seasonal and multi-year cycles can affect climate conditions in regions

D. That climate does naturally vary, but the change is not necessarily uniform

E. How paleoclimate records can help scientists understand past climate patterns and processes

F. How recent changes in climate cannot be explained by natural processes and evidence points to human impact on the system

G. That carbon dioxide and greenhouse gases released into the atmosphere from human activities will remain in the atmosphere for centuries or longer

Conflating weather and climate is common and understandable, especially when climate is sometimes defined as averaged weather. But such definitions obscure the fact that processes driving weather, which often are driven by convection on a local or regional scale over the course of hours or days, differ from the processes that drive climate, which is influenced by the annual cycle of the Earth orbiting the Sun and the resulting seasonal changes as well as longer-term cycles and processes. In elementary grades, short-time scales (daily, seasonal, and annual) set the foundation for longer-term, more complex processes.

Missed Conceptions

- Climate is related to but not the same as weather and visa versa: Weather occurs over minutes, hours, and days; climate occurs over seasons, years, decades, and beyond

- Temporal scaling using log10, powers-of-ten is how scientists are able to collapse time and more easily study processes at different time scales

- How paleoclimate research is conducted, its value and margins of error

- Rates of change (abrupt change may be a matter of hours with weather or over seasons, years, or decades in terms of climate)

Doing the Math

This principle is particularly rich with opportunities to weave together numeracy with climate science, setting the stage for Principle 5, which is focused on how scientists know what they know and the ways they observe, model, and communicate their findings about the climate system. In particular, this principle, with its focus on temporal and spatial scales, lends itself to teaching about decimals, exponents, conversions, and particular measures, such as carbon emissions.

CLEAN Recommends

Middle School

> Changes Close to Home—Smithsonian National Museum of Natural History
>> http://cleanet.org/resources/41792.html

High School

> Exploring Regional Differences in Climate Change—Denise Blaha and Rita Freuder; Earth Exploration Toolbook from TERC
>> http://cleanet.org/resources/41852.html

> Normal Climate Patterns—Betsy Youngman and LuAnn Dahlman; Earthlabs from TERC
>> http://cleanet.org/resources/41838.html

Introductory College

> Carbon Dioxide Exercise—Randy Richardson, SERC—Starting Point Collection
>> http://cleanet.org/resources/41805.html

Upper-Level College

Global Temperatures—Robert MacKay, SERC Starting Point and Columbia University Earth and Environmental Science Faculty

http://cleanet.org/resources/41830.html

TELLING TIME: A Brief History of the Universe in Powers-of-Ten

The measurement of time, known in the scientific realm as temporal scaling, adds a vital dimension—specifically the fourth dimension—to mastery of the science, allowing for the measure of rates of change over time.

While weather is measured on time scales of hours and days, climatologists often employ logarithmic temporal scaling using the annual cycle as the reference point and then using an exponential "powers-of-ten" framework to examine the processes and rates of change over annual to decadal (1 to 10 years), decadal to centennial (10 to 100 year) scales, and so forth. One of the NGSS performance expectations at the high school level Earth Science, HS-ESS2-4, calls for students to demonstrate their mastery of climate processes at various time scales. More will be said about this later, but pedagogically, it makes sense to start with the basics—such as seasonal changes over the course of a year—and build on prior knowledge to help learners connect the dots between concepts.

For example, students are asked to observe and measure weather phenomenon, such as precipitation, and then examine how precipitation patterns change over different time scales over the course of a year in accordance to the seasons, with weather occurring over short-time periods—hours and days—while seasons occur over weeks and months. Ultimately climate picks up where weather leaves off, spanning months, years, decades, and beyond.

The NOAA Climate TimeLine tool, (http://www .ncdc.noaa.gov/paleo/ctl), which in the spirit of full disclosure I led the development of more than a decade ago while at the NOAA Paleoclimatology Program, uses the powers-of-ten approach to collapsing the climatic processes and past history down to demonstrate how this logarithmic approach is used by scientists to study the processes and rates of change involved.

Using the frame of powers-of-ten, we can see the climate of the universe can change abruptly, as in the case of its initial birth, or slowly, as elements gradually gathered mass, forming galaxies and solar systems amidst the vast, expanding emptiness of space.

The age of the Earth, "third stone from the sun" as it has been described, is estimated to be 4.54 billion years. The universe is between 13.77 and 13.82 billion years old; given the fact that scientists studying the age of the universe are using the length of time it takes a particular planet now to circle its solar center more than nine billion years before that planet even existed should give pause . . . and a little wiggle room for rounding errors.

In the beginning, according to well-established scientific thought known as the Big Bang Theory—obviously not to be confused with the popular television series—things happened on an extremely short-time scale: From out of nowhere and ultra quickly by any measure (10^{-37} seconds), an infinite density and temperature expanded and began to cool. Over time the early elements of hydrogen then helium gave way to the building blocks of the cosmos and eventually a habitable Earth: carbon, nitrogen, oxygen, iron, and later heavier and heavier elements, the most recent being human synthesized, like plutonium.

Once formed, geologic processes over hundreds of millions of years have shaped the Earth's surface. Liquid water provided the medium for biologic systems to begin to take root and establish themselves roughly a billion years after the early Earth solidified, and the atmospheric, magnetic, and gravitational dynamics of the planet allowed microbial life to take hold and begin to alter the atmosphere. The invention of photosynthesis altered the atmosphere and thereby the climate over time (Kiang, 2008), first with bacteria that produced sulfur or sulfate rather than oxygen and later, less than two billion years after the formation of the planet, with cyanobacteria and eventually algae that began releasing oxygen into the atmosphere, substantially changing the chemistry and thus the climate of the planet.

During the hundreds of millions of years of evolution of the organic, carbon-based variety (as opposed to the more cosmological evolution of star systems and planets), climate has altered over long timescales, as plate tectonics and geological forces interacted with orbital and solar dynamics to relatively slowly change the Earth's climate. It has also altered fairly rapidly, with five mass extinctions caused at least in part by changes in climate, leading to the demise of numerous species.

As the logarithmic chart inspired by Mitchell's 1976 overview of climatic variability depicts on the Climate Timeline website (http://www.ncdc.noaa.gov/paleo/ctl/about1.html), major climate processes caused by variations in the Earth's orbit—Milankovitch cycles—begin at the 100,000 time scale, with other important natural processes, such as El Nino Southern Oscillation occurring at the decadal (century to 10-year) scale.

While our human ancestors date back millions of years, Homo sapiens began to gradually migrate out of Africa less than 100,000 years ago, when ice sheets still covered much of Europe, Asia, and North America. At the end of the last Ice Age just over 10,000 years ago, agricultural societies began to take root. And the rest, as they say, is history, which is often broken into significant centuries, decades, or years.

The annual cycle and resulting seasonal changes is often overlooked and underappreciated as a key process driving climate variations, energy consumption, and the behavior and responses of organisms, such as growth and decay of plants and the migration and hibernation of some animals. Finally, at the daily or diurnal cycle of 24 hours, which does not fit neatly into the "powers of 10" exponential approach to years, being 0.0027 of a year, weather and processes like photosynthesis are driven by day and night.

5. OUR UNDERSTANDING OF THE CLIMATE SYSTEM IS IMPROVED THROUGH OBSERVATIONS, THEORETICAL STUDIES, AND MODELING.

The five concepts of the fifth principle of *Climate Literacy* emphasize the following:

A. That the climate system can be understood by the same tools of science as the rest of the Universe

B. How observations from a wide range of instruments are the foundation of our understanding of the climate system

C. How computer models, which are iterated to improve accuracy, help in understanding the complex interactions of the climate system

D. How weather forecasting differs from climate projections

E. That current research and projections of climate scientists are sufficiently accurate and robust enough to help humans evaluate options and responses

Two key questions in the minds of many educators and learners alike are, how do scientists know what they know? And, how certain are they that climate is changing and that humans are responsible? Good science educators have long used the process of inquiry to allow students to experience the process of science from the inside out, and this principle is designed to help demystify how climate scientists know what they know . . . and don't know, since there are some areas, such as the rate of speed that sea level will rise and ice sheets will melt in various locations, that are still unknown. Because the NGSS have students learning inquiry skills through observation, analysis, modeling, and developing and testing engineering solutions throughout the grade bands, by middle and certainly high school and into college, they will be able to gain strong climate science literacy skills.

Missed Conceptions

- While the term *theory* may suggest it is something unproven, in fact it means that it is a well-substantiated explanation, like the "theory of gravity;" to claim that "global warming is just a theory" is to take advantage of (or fall victim to) this confusion

- Assuming that weather forecasting and modeling is basically the same as climate projections, when in fact they involve different physical processes and modeling assumptions

- Not appreciating the robustness and self-correcting nature of the scientific peer-review process, from data collection, analysis, modeling, review, and publication

Doing the Math

An element of confusion associated with this principle is the concept of uncertainty. In common parlance it means being doubtful and unsure, but in science and mathematics, uncertainty refers to the range of possible statistical error because of the instruments or equations used that need to be considered when looking at the data. Scientists are used to and expect error bars around their measurements and projections and often start their research by leading with their uncertainties. But when conveying to nontechnical audiences their research findings, putting uncertainties up front conveys a sense of doubt and a lack of confidence, even though the confidence level may be very strong.

The IPCC Fourth Assessment used ten gradations of probabilistic likelihood ranging from virtually certain (> 99%) to exceptionally unlikely (< 1%), but climate communicator Susan Joy Hassol has suggested using an approach more like the weather service when predicting precipitation: 20%—slight chance, 30–50%—chance, 60–70%—likely, and 80–100%—it's going to rain or snow. Below 10% or above 80% should have no qualifying terms because of their higher confidence levels.

Making sure learners appreciate how scientific certainty and uncertainty differs from common usage of the terms is an important step toward improving data literacy.

CLEAN Recommends

Middle School

> Using a Very, Very Simple Climate Model in the Classroom—Randy Russell and Lisa Gardiner, Windows to the Universe
>
> > http://cleanet.org/resources/41874.html

High School

> U.S. Historical Climate: Excel Statistical—R. M. MacKay, SERC Starting Point
>
> > http://cleanet.org/resources/41823.html

Introductory College

> Climate and Civilization: The Maya Example—Katherine Ellins, Jeri Rodgers, and James Cano, Texas Earth and Space Science Revolution
>
> > http://cleanet.org/resources/41875.html

Upper-Level College

> Global Temperatures—Robert MacKay, SERC Starting Point and Columbia University Earth and Environmental Science Faculty
>
> > http://cleanet.org/resources/41830.html

6. HUMAN ACTIVITIES ARE IMPACTING THE CLIMATE SYSTEM.

The five concepts that round out the sixth principle focus on the following:

A. The solid agreement of scientific studies indicates human activities are responsible for increases in greenhouse gases and thus the recent global temperature increases.

B. The burning of fossil fuels, which result in greenhouse gases that can remain in the atmosphere for hundreds of years, is the primary reason for the global temperature increases.

C. This and other human activities, including changes of land cover, alter the balance of the climate system.

D. The impact of these activities is reduced biodiversity and resilience.

E. If the warming exceeds 2 to 3 °C (3.6–5.4 °F) in the next 100 years, negative impacts are likely outweigh positive ones.

Beginning in kindergarten, learners are considering how humans (and other organisms, but especially humans) alter their environment, and they consider ways to minimize negative impacts through appropriate, effective responses. Officially, human-induced anthropogenic climate change isn't delved into until middle school, but it is likely many students will have a general idea of what's what by then. Drilling into the details with authentic data and real-world scenarios can help bring home the fact that we do have evidence to attribute changes to the climate system and, indeed, other global changes, which are happening at a fast pace, to human activities—primarily, but not exclusively, the burning of fossil fuels that increase greenhouse gas concentrations, thereby heating the planet.

Missed Conceptions

- There are multiple lines of evidence attributing current global climate change to human activities.

- There is robust agreement among climate scientists that human activities, primarily burning fossil fuels, are altering the climate system for centuries to come.

- Reducing and stabilizing greenhouse gas emissions will not stop climate change, but may slow the rate of change and to some extent minimize impacts.

- Related human activities are causing changes in ecosystems, such as habitat and ecosystems destruction.

Doing the Math

This principle offers a strong link to *Energy Literacy*, including the measure of energy generated by carbon-based fuels and the reduction of wasted energy through conservation and energy efficiency. There is no shortage of opportunities to quantify, analyze, and model human impacts on the environment and our use of energy, beginning in the elementary grades with basic arithmetic, and move into increasingly sophisticated data collection and analysis, drawing upon skills of geometry and algebra and ultimately calculus and computer modeling.

CLEAN Recommends

Middle School

Automotive Emissions and the Greenhouse Effect—Texas State Energy Conservation Office

http://cleanet.org/resources/41915.html

High School

Mauna Loa CO2 Collection Data—Connecticut Energy Education

http://cleanet.org/resources/41907.html

Introductory College

Global Climate Change: The Effects of Global Warming—Teachers' Domain

http://cleanet.org/resources/41876.html

Upper-Level College

Using a mass balance model to understand carbon dioxide and its connection to global warming—Robert MacKay, SERC—Teaching Quantitative Skills in Geoscience Collection

http://cleanet.org/resources/41868.html

7. CLIMATE CHANGE WILL HAVE CONSEQUENCES FOR THE EARTH SYSTEM AND HUMAN LIVES.

Under Principle 7 there are six concepts that describe the key ways that climate change is already and will continue to influence ecosystems and society:

A. Sea level is rising along many coasts, damaging communities, increasing risk of storms, and contaminating fresh water.

B. Changes in precipitation and temperature alter freshwater distribution, with regions dependent on snowpack and mountain glaciers being impacted.

C. Extreme weather is projected to increase, with increased heat waves, more intense precipitation events, and increased droughts.

D. The ocean is becoming acidified as it absorbs carbon dioxide, decreasing the pH.

E. Terrestrial and marine ecosystems are being altered, changing disease and species distribution patterns.

F. Human health will be affected in a wide range of ways, from experiencing more climate-sensitive infectious diseases, to changing crop yields and degraded air and water quality.

The fact that these concepts are serious and are often avoided because of their depressing message is a particular challenge for educators, especially for those of younger students. They also point to the importance of circling back to and infusing the "Guiding Principle"—humans can take actions to reduce climate change and its impacts—throughout the process. But to reduce impacts and prepare for changes already occurring, it is important to know the essentials. Many professionals in the military, in water and public works departments, in resource management, public health, insurance companies, and corporations large and small are grappling with how to ready themselves for what military planners call threat multipliers while at the same time looking for opportunities to reduce waste and streamline processes.

Missed Conceptions

- Climate change is something distant and won't affect me; in fact the consequences of climate change are already occurring, in our front and back yards.

- Someone else is causing the problem and there's nothing much I can really do. While climate and energy issues can be overwhelming, there is much that can be done to understand the causes, effects, scope, and scale of the challenges in order to troubleshoot them and develop effective responses.

Doing the Math

Measuring the impacts of global change can be daunting and depressing, but it is vital to understand what we can and be clear about what is unknown and uncertain. Using mathematics to measure changes brings a real-world dimension for students to apply their skills to the analysis of topics such as sea level rise, changes in ocean pH, or the probabilities of intensified precipitation events or health risks. Once we are past the aversion to facing what has been described as the sixth great extinction, we can access fascinating, inspiring insights into inner workings of the world and the potential for transforming the human dimension to be more climate smart, energy wise, and ready for the future, come what may.

CLEAN Recommends

Middle School

> Impacts of Topography on Sea Level Change—Lise Whitfield, Bill McMillon, Judy Scotchmoor, and Phil Stoffer, DLESE (Digital Library for Earth System Education)
>
> http://cleanet.org/resources/41831.html

High School

> Off Base—Acidity of oceans—NOAA Ocean Explorer
>
> http://cleanet.org/resources/41828.html

Introductory College

> Google Earth Tours of Glacier Change—Mauri Pelto, SERC—On the Cutting Edge Collection
>
> http://cleanet.org/resources/41855.html

Upper-Level College

> Exploring the Link between Hurricanes and Climate using GCM Results—Cindy Shellito, SERC—On the Cutting Edge Collection
>
> http://cleanet.org/resources/41853.html

In Chapter 5, the focus is on *Energy Literacy* principles, the other half of the equation in teaching and learning about climate and energy science.

ADDITIONAL RESOURCES

Additional information from the National Science Teachers Association that highlight weather and climate-related disciplinary core ideas and related resources include the following:

Elementary

> standards.nsta.org/DisplayStandard.aspx?view=topic&id=3
>
> standards.nsta.org/DisplayStandard.aspx?view=topic&id=14
>
> standards.nsta.org/DisplayStandard.aspx?view=topic&id=21

Middle School

> standards.nsta.org/DisplayStandard.aspx?view=topic&id=37
>
> standards.nsta.org/DisplayStandard.aspx?view=topic&id=38

High School

> standards.nsta.org/DisplayStandard.aspx?view=topic&id=53
>
> standards.nsta.org/DisplayStandard.aspx?view=topic&id=54

CHAPTER 5

Teaching Energy Literacy

A Dynamic Duo

Energy and climate literacy are inherently complementary and help lay a solid foundation for understanding and responding to global change. Energy concepts are integral to understanding climate processes and the causes of climate disruption, and climate change provides the context for why alternatives to our reliance on fossil fuel energy sources are important. Being energy aware and efficient and understanding the essentials of energy in our lives goes beyond climate literacy and can help us make informed choices in many aspects of our lives.

Learning about energy and its role in the universe in general and in our lives in particular is a lifelong endeavor. Yet energy is often taught in an ad hoc, piecemeal way, rarely treated as a cross-disciplinary, integrating theme for learning. Like climate, energy crosses many disciplines and may fall through the cracks of traditional disciplines. Even more than climate, the study of energy is an opportunity to learn and apply knowledge to real-world, everyday experiences and challenges. In many respects, energy is the über theme that is often considered to be primarily the domain of physical science if not physics but in fact comes into play as the invisible (since we can't exactly see energy but only its potential or effects) factor driving not only our cars and our lives but also the climate, ecosystems, and indeed, the entire universe in which we are embedded.

Energy is, by definition, not stuff. Stuff is substantive, made of matter. Although modern physics steers away from the term *matter*, preferring to focus on the equivalence of mass and energy conveyed in Einstein's $E = mc^2$ equation—energy equals mass times the speed of light squared—it can be initially helpful to accept a common definition of matter as an object that has mass and volume. And while some scientists may cringe at describing the "flow" of energy, the metaphor can be a good entry point for deeper investigations of energy in all its nuance and complexity.

Energy, technically, is no-thing, although all things have potential and embedded energy in them. In some respects, energy is a riddle. It is not easy to teach or learn about, especially if the emphasis is placed on the dry, technical aspects and equations. $F = ma$, or force equals mass times acceleration, is just an equation until it can be brought to life in the mind of the learner. Emphasizing the relationship between energy and the mass of an object or system can help bring it from the abstract to the everyday. We might even think of it as a Zen koan to contemplate and wrestle with: how does energy matter? Or a riddle. How much does light weigh? (Answer: It is very light!)

Once the door into the realm of energy is open, it becomes inherently exciting and can be learned in a way that is immersive and interesting. Energy cannot be directly observed but can be calculated from its state as a property of a physical system—flowing through matter and systems over time, in a sense evolving from its most concentrated state at the beginning of time as the universe evolves and expands over time. And while energy cannot be created or destroyed, it is in a sense eroding, losing its steam, so to speak, as entropy fragments and dissipates energy from concentrated sources into the wider (and ever-expanding) universe, leading some theoretical physicists to anticipate that the universe will eventually burn out and run down.

To help unpack the proverbial nuts and bolts of energy, we turn to the *Energy Literacy* framework. Modeled on the *Climate Literacy* framework and developed under the leadership of the U.S. Department of Energy through a series of meetings and workshops, this framework defines an energy literate person as someone who

- Can trace energy flows and think in terms of energy systems

- Knows how much energy he or she uses, for what, and where the energy comes from

- Can assess the credibility of information about energy

- Can communicate about energy and energy use in meaningful ways

- Is able to make informed energy and energy use decisions based on an understanding of impacts and consequences

- Continues to learn about energy through his or her life

STUDYING SPAGHETTI STRANDS

Human energy consumption is naturally a key element of energy literacy, and certainly by graduation from high school, learners should be able to make sense of a spaghetti or Sankey diagram describing energy generation, consumption, and waste, called "unused," but in reality, usually heat vented into the environment, creating thermal pollution. At first glance, such diagrams are overwhelming, a confusing technical chart of arrows pointing every which way. The mathematics implied— quads of BTU or British Thermal Units—are daunting to any novice. But on closer examination, each arrow tells a story of the life cycle of an energy source, whether renewable, nuclear, or fossil fuel; the use of the energy, such as electricity; or energy services, specifically residential, commercial, industrial and transportation.

The Sankey diagram in Figure 5.1 is from the Lawrence Livermore National Laboratory, and helps to demonstrate the degree of efficiency in residential, transportation, and commercial sectors, with industrial being substantially more efficient than the transportation sector,

(Continued)

while making it clear that there is room for improvement in all areas.

To fully appreciate energy in our lives, it may be helpful to step back and examine energy in general rather than diving headfirst into trying to untangle the spaghetti strands of a Sankey. Like climate, energy is often ignored, misunderstood, or skimmed over in school. Some educators focus on energy but ignore climate, but both, linked together, are vital for preparing learners for the 21st century. By connecting energy and climate as integrating themes, it is possible to "close the loop" of the carbon cycle and the environmental and climatic impacts of fossil fuel consumption.

This chapter will be nearly identical in approach and format to the chapter on climate literacy. The table on pages 82–83 provides an initial overview of how energy concepts (presented here in an abbreviated form) relate with disciplinary core ideas identified in the Framework for K–12 Science Education and the related Performance Expectations described in the NGSS. Summaries of the concepts are shortened from the original literacy document. The text of the *Energy Literacy* guide is available in Appendix III.

As with the chapter on climate literacy, we'll focus instead on the essential principles and their supporting concepts, followed by Missed Conceptions and Do the Math sections. A CLEAN Recommends area will also be featured.

1. ENERGY IS A PHYSICAL QUANTITY THAT FOLLOWS PRECISE NATURAL LAWS.

Within the first principle are related concepts that help define what energy is (and isn't), including the following:

1. Energy is transferred from system to system, allowing work to occur.

2. Heat is transferred between systems when the temperature or thermal energy is different through a process of convection, conduction, or radiation.

3. The amount of energy into a system equals the amount leaving the system, since energy is neither created nor destroyed.

4. Energy can be lost to the surroundings when energy is transferred from system to system, with less being lost the more efficient the system is.

5. Energy can be categorized into forms, such as light, chemical, or elastic energy, with all energy being either kinetic (with motion) or potential.

6. Chemical and nuclear reactions transfer and transform energy on different scales.

7. Energy is measured with a variety of units (joules, calories, BTUs, kilowatt-hours) and the units of energy can be converted to other units.

8. The rate or measure of energy transferred between systems over time is called power, with one joule per second equaling one watt.

FIGURE 5.1 Estimated U.S. Energy Use in 2013: ~97.4 Quads

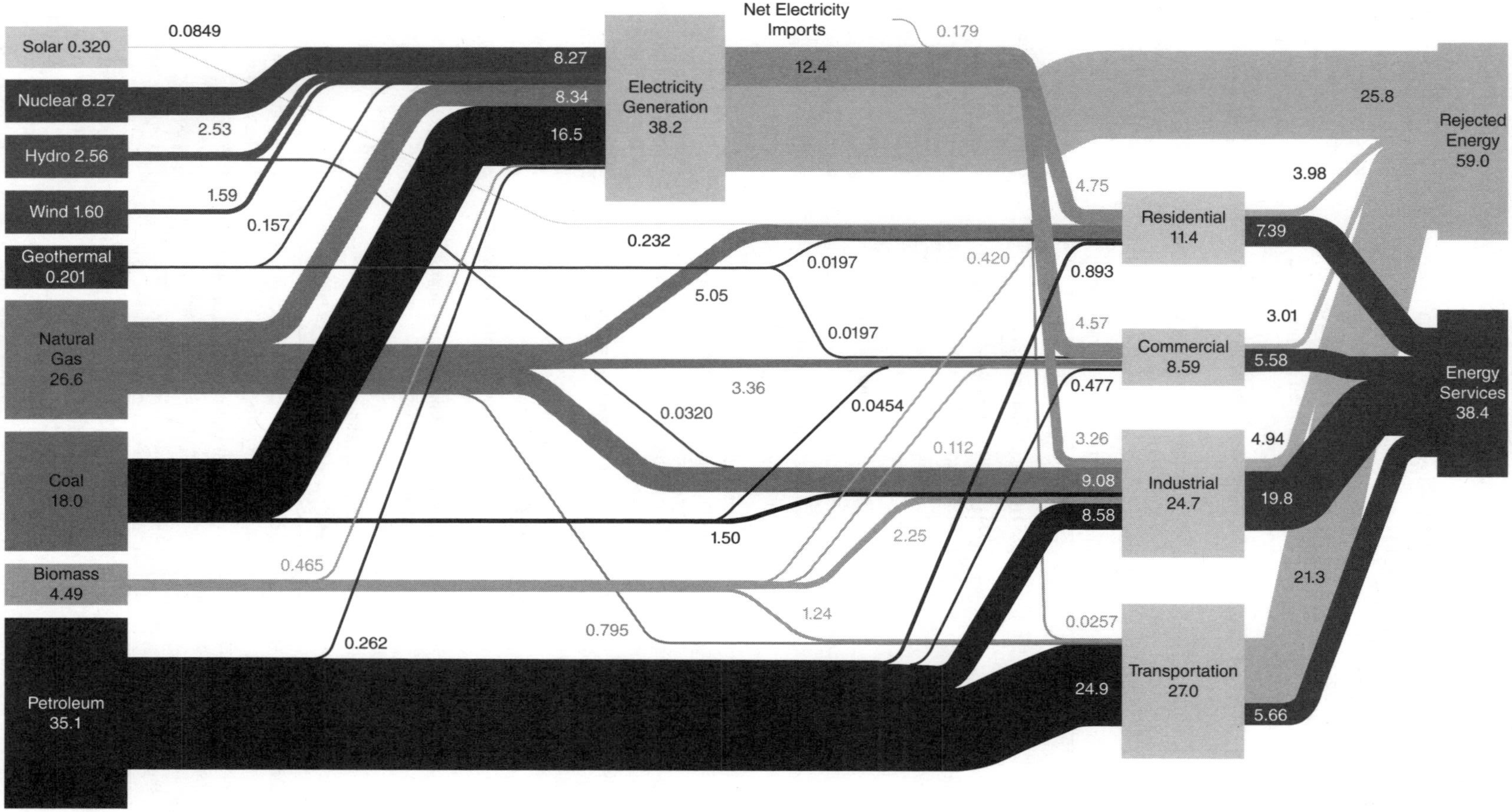

SOURCE: Lawrence Livermore National Laboratory. https://flowcharts.llnl.gov

ESSENTIAL PRINCIPLES	VITAL CONCEPTS (IN BRIEF)	DISCIPLINARY CORE IDEAS FROM THE NRC FRAMEWORK FOR K-12 SCIENCE EDUCATION	NGSS PERFORMANCE EXPECTATIONS
Energy Literacy 1. Energy is a physical quantity that follows precise natural laws.	Definitions and types of energy; thermodynamics; transfer and transformation of energy in matter; ways to measure; power as a measure of energy transfer rate.	PS1.A: Structure of matter PS1.B: Chemical reactions PS2.A: Forces and motion PS2.B: Types of interactions PS3.A: Definitions of energy PS3.B: Conservation of energy and energy transfer PS3.C: Relationship between energy and forces PS3.D: Energy in chemical processes and everyday life PS4.A: Wave properties PS4.B: Electromagnetic radiation	From K-PS2 (effects of pushes and pulls on motion, speed and direction of object) and K-PS3 Energy (sunlight on Earth's surface), to 4-PS3 Energy (relating speed to energy, transfer of energy, changes in energy, converting energy), MS-PS3 Energy (relationships among energy, matter), and all HS-PS (all high school physical science-related topics).
2. Physical processes on Earth are the result of energy flow through the Earth system.	Physical processes; Sun drives most weather, climate, and hydrologic cycle; greenhouse gases affect flow; lag time responses.	ESS1.A: The universe and its stars ESS1.B: Earth and the solar system ESS1.C: The history of planet Earth ESS2.A: Earth's materials and systems ESS2.C: The roles of water in Earth's surface process ESS2.D: Weather and climate ESS2.E: Biogeology ESS3.A: Natural resources ESS3.C: Human impacts on Earth system ESS3.D: Global climate change PS4.B: Electromagnetic radiation	From K-PS3-1 & 2 (effect of sunlight on Earth's surface and building a structure to reduce warming effect of sunlight), to MS-ESS2 Earth's Systems (cycling of materials and flow of energy), and all HS ESS Earth and Space Science.
3. Biological processes depend on energy flow through the Earth system	Sun as major energy source for organisms (photosynthesis); entropy in food chain; food webs and ecosystems affected by changes in energy and matter; humans influence energy flows.	PS3.D: Energy in chemical processes and everyday life L21.B: Growth and development of organisms LS1.C: Organization of matter and energy flow in organisms LS2.A: Interdependent relationships in ecosystems LS2.B: Cycles of matter and energy transfer in ecosystems LS2.C: Ecosystem dynamics, functioning, and resilience LS4.C: Adaptation	From K-LS1 From Molecules to Organisms (patterns of what plants and animals, including humans need to survive), LS2 Ecosystems: Interactions, Energy, and Dynamics (role of sunlight on plant growth), MS-LS2 Ecosystems: Interactions, Energy, and Dynamics (cycling of matter and flow of energy in ecosystems), and all HS-LS Life Science.
4. Various sources of energy can be used to power human activities, and often this energy must be transferred from source to destination.	Sources of energy for human consumption; limits and constraints; fossil fuels as buried solar energy; transport, generation, storage of energy and related benefits and drawbacks.	PS3.D: Energy in chemical processes and everyday life LS1.C: Organization for matter and energy flow in organisms LS2.B: Cycles of matter and energy transfer in ecosystems ES2.A: Earth's materials and systems ESS3.A: Natural resources ESS3.C: Human impacts on Earth systems	Primarily from 4-ESS3 Earth and Human Activity (energy and fuels, impact on environment, reducing impact), MS-ESS3 Earth and Human Activity (resources, population, and consumption), HS-HSS3 Earth and Human Activity, but also an opportunity to bring in life sciences by focusing on food web.

ESSENTIAL PRINCIPLES	VITAL CONCEPTS (IN BRIEF)	DISCIPLINARY CORE IDEAS FROM THE NRC FRAMEWORK FOR K-12 SCIENCE EDUCATION	NGSS PERFORMANCE EXPECTATIONS
5. Energy decisions are influenced by economic, political, environmental, and social factors.	Levels of energy decision-making and systems-based approach; infrastructure/technology inertia; political, environmental, and social factors of decisions.	ESS3.A: Natural resources ESS3.B: Natural hazards ESS3.C: Human impacts on Earth systems ESS3.D: Global climate change All eight of the practices of science and engineering relate to decision-making and should be emphasized.	From 5-ESS3 Earth and Human Activity (how communities use science to protect resources and environment), to MS-ETS1 Engineering Design (constraints on design solutions), HS-ESS3 Earth and Human Activity (cost-benefit analysis of resources, etc.), and HS-ETS1 Engineering Design (criteria and constraints for solutions).
6. The amount of energy used by human society depends on many factors.	Conservation of energy (physical law) vs. conserving energy; increased human demand of energy; ways of decreasing use (per capita, total); energy (amount and quality) embedded in products and services; monitoring and calculating energy usage.	PS3.A: Definitions of energy PS3.B: Conservation of energy and energy transfer PS4.C: Information technologies and instrumentation LS2.B: Cycles of matter and energy transfer in ecosystems ESS2.D: Weather and climate ESS3.A: Natural resources ESS3.B: Natural hazards ESS3.C: Human impacts on Earth system ESS3.D: Global climate change All eight of the practices of science and engineering relate to decision-making and should be emphasized.	From 5-ESS3 Earth and Human Activity (how communities use science to protect resources and environment) to MS-ETS1 Engineering Design (constraints on design solutions), HS-ESS3 Earth and Human Activity (cost-benefit analysis of resources, etc.), and HS-ETS1 Engineering Design (criteria and constraints for solutions).
7. The quality of life of individuals and societies is affected by energy choices.	Economic, national, and environmental security impacts of energy choices; access and limits of fossil fuel challenges; how energy access affects quality of life; impact of energy decisions or access on vulnerable populations.	ESS2.A: Earth's materials and systems ESS3.A: Natural resources ESS3.C: Human impacts on Earth system ESS3.D: Global Climate Change All eight of the practices of science and engineering relate to decision-making and should be emphasized.	From 5-ESS3 Earth and Human Activity (how communities use science to protect resources and environment), to MS-ETS1 Engineering Design (constraints on design solutions) HS-ESS3 Earth and Human Activity (cost-benefit analysis of resources, etc.), and HS-ETS1 Engineering Design (criteria and constraints for solutions).

In the NGSS, the foundation for energy literate individuals is established in kindergarten, where learners observe how energy from the Sun affects the Earth's surface, look at the forces and interactions between objects, explore what plants and animals need to survive, note changes in weather patterns over time, and learn about ways humans can minimize their impact on the local environment. Over the next few years, these themes are further explored, and in fourth grade, with PS3 Energy, there is much more detailed analysis of energy, how it is derived from fuels, and what the impact is on the environment and society. Ideally, students could add to their knowledge through field trips to local energy-generating facilities, but these trips have become increasingly difficult to arrange for a variety of reasons. Online virtual tours, even if available, are not the same. Energy as a crosscutting theme in the NGSS continues through middle and high school, with the focus increasingly on engineering, technological, and societal responses to the challenge of balancing energy needs and demand with the environmental and socioeconomic impacts and constraints associated with energy consumption.

It is important to be able to differentiate between the different types of energy, but the definitions may not stick with learners because they are sometimes conveyed in a dry, technical tone that learners may have a difficult time relating to. The definitions of key energy types learners should know include the following:

- Mechanical energy (the sum of kinetic and potential energy)
  - o Gravitational potential energy (work required to elevate objects against gravity)
  - o Kinetic energy (energy of motion)
  - o Elastic potential energy (mechanical energy stored in object, released through distortion of shape or volume)
- Thermal energy (temperature of system)
- Radiant energy (energy of electromagnetic waves)
- Chemical energy (energy released or absorbed through chemical reaction)
- Nuclear energy (energy released through fusion, fission, or radioactive decay)

Missed Conceptions

- *Energy is an object—such as an electrical power plant or transmission line, oil, gas, coal, a wind turbine, or a solar panel.* Confusion between energy and a source of energy is common. Being clear about the language is important.

- *Energy is too complicated and technical to understand; there's too much math.* Although some understanding of energy can involve math, many basic concepts can be accessible without daunting equations. And the context of energy in the Earth system and our everyday lives can help make the mathematics more relevant to the learners.

Doing the Math

Energy can indeed be measured because it abides by laws of thermodynamics and other well-established scientific principles. To measure the universe, whether at the micro or macro scale, science relies primarily on the seven SI base units as determined by the Système Internationale d'Unités (SI) (International Bureau of Weights, n.d.), which in brief measure the following:

1. Length (meter)
2. Mass (kilogram)

3. Time (seconds)

4. Electric current (amper)

5. Temperature (kelvin)

6. Luminous intensity (candela)

7. Amount of substance (mole)

Defining energy requires the first three, and measuring it often uses the following three. (The seventh unit, the mole, is not a true metric for measure, as it is a way to convey amounts of chemical substances.)

Measuring energy, whether potential or kinetic, can begin in kindergarten, when mostly qualitative measures of sunlight and the motion of objects are initiated and inquiry into the world of energy begins in earnest. Clearly, explorations of energy are naturally synergistic with mathematics, and educators should look for ways to weave together appropriate learning opportunities to maximize teachable moments.

CLEAN Recommends

Middle School

Energy flows—NEED Project—Putting Energy into Education
http://cleanet.org/resources/41911.html

High School

Energy Unit Conversion Calculator—U.S. Energy Information Administration
http://cleanet.org/resources/42687.html

College

Solar Water Heater—Landon B. Gennetten, Lauren Cooper, Malinda Schaefer Zarske, and Denise W. Carlson, Teach Engineering from Integrated Teaching and Learning Program
http://cleanet.org/resources/41891.html

Power Metering—Department of Energy Academies Creating Teacher Scientists, Department of Energy
http://cleanet.org/resources/43009.html

2. PHYSICAL PROCESSES ON EARTH ARE THE RESULT OF ENERGY FLOW THROUGH THE EARTH SYSTEM.

In investigating this second principle, key related concepts include insights about the Earth system, which reveal the following:

1. Energy is changing because of the flow of energy through living and nonliving systems.

2. Sunlight and tidal energy are from external sources of energy, while radioactive isotopes and gravity, along with Earth's rotation, are internal sources.

3. Weather and climate are primarily driven by the Sun, with unequal warming influencing the Earth system.

4. Water's unique qualities are integral to the storage and transfer of energy, with the Sun propelling the hydrologic cycle.

5. Energy drives the flow of matter and processes, such as the carbon cycle.

6. Greenhouse gases, like carbon dioxide and water vapor, play a major role in the temperature of the atmosphere because of their ability to absorb outgoing infrared heat from the Earth.

7. Changes in the Earth's energy balance may be delayed for years, even decades, because of time lag or thermal response.

This second energy literacy principle overlaps substantially with many of the *Climate Literacy* principles, particularly the first two: The Sun is the primary source of energy for the Earth's climate system, and climate is regulated by complex interactions among components of the Earth system.

One of several reasons why these vital elements of climate and energy were often missing or skimmed over in traditional science courses is that this content and the related analytical skills traditionally fall within the realm of Earth science. Assuming students have taken a course in Earth science at all (which may not be required in some states), they may have had it in middle school, with an emphasis on geological processes, rocks, and minerals. High school courses where these concepts are taught are often for students not planning to attend college, so ironically students heading to college, often taking AP biology and chemistry courses, may be less likely to learn about climate change than students who have less modest plans after high school graduation.

The NGSS could change this situation since they are aimed at all students. The NGSS approach of teaching climate- and energy-related content and skills makes simple sense: Beginning in kindergarten and continuing through high school and beyond, focus on skills building, problem solving, scaffolded learning, and building upon prior learning progressions in order to foster true science literacy.

For example, the water cycle, driven primarily by energy from the Sun and gravity, is introduced in primary grades. But in succeeding grade levels there is learning occurring around the never-ending nuances and complexity of water, from its molecular structure and infrared-absorbing qualities and the conditions on the planet that allow it to exist in solid, liquid, and vapor phases (even, briefly, simultaneously at the triple point around 0.01 °C on Earth's surface), to the ways that humans have harnessed, domesticated, and contaminated the water of the world.

The ways energy drives the carbon cycle, particularly the capture and storage of energy through photosynthesis, the role of carbon dioxide and methane in the greenhouse effect, the formation of fossil fuels, and the long-term storage of carbon in sedimentary deposits like limestone, can also be unpacked when teaching concepts relating to this principle.

Missed Conceptions

- *Water vapor is the most abundant greenhouse gas and therefore we don't need to pay attention to the other gases.* Yes, H_2O is the most abundant greenhouse gas, but its residence time in the atmosphere is on the order of hours and days, with a nine-day average residence time rather than decades and centuries, as is the case of many other greenhouse gases.

- *If we cut back on greenhouse gas emissions, the climate will quickly stabilize to a safe climate.* Actually, greenhouse gases in the atmosphere with long residence times and delayed thermal response mean that some warming is already in the proverbial pipeline.

- *Carbon and carbon dioxide are bad . . . or they are food for plants and therefore not a problem.* For those who have never had a course in the essentials of climate change, it is easy to get the impression, listening to statements about carbon pollution—which stress cutting back carbon dioxide emissions that result from our energy consumption or emphasize reducing the size of carbon footprints—that carbon dioxide is an evil that should be eradicated from the planet. Equally as confusing can be statements that carbon dioxide is wonderful and without any negative attributes because plants need it for photosynthesis. In a sense, both are partly true—human activities are releasing forms of carbon into the atmosphere that are altering the climate and increasing heat levels, and plants do in fact need carbon dioxide from the air to grow and build their mass. But understanding the cycle of carbon fully trumps incomplete or misleading bits of information.

Doing the Math

Measuring inputs and outputs of the Earth system is exactly what early climate scientists like Joseph Fourier and later John Tyndall and Svante Arrhenius did in the 19th century, without the benefit of supercomputers or satellites. Today, paleo data from the geologic record, including fossils, ice cores, tree rings, and even packrat dens, afford us glimpses of past changes in climate and the environment. Knowledge of energy flows through the Earth system aids scientists to numerically model projected future changes in the Earth's energy balance and, thereby, its climate and environmental systems.

CLEAN Recommends

Middle School

> Global Energy Balance—Scripps Institution of Oceanography
> http://cleanet.org/resources/42819.html

> The Carbon Cycle—Maree Lucas, West Virginia University Online
> http://cleanet.org/resources/43440.html

High School

> Earth as a System—WGBH/Boston
> http://cleanet.org/resources/43013.html

> Star Power!—NASA Discovery Program, Carnegie Institution of Washington and Johns Hopkins University Applied Physics Laboratory
> http://cleanet.org/resources/42694.html

> Striking a Solar Balance—NASA, nasa.gov/multimedia
> http://cleanet.org/resources/43154.html

College

Modeling Earth's Energy Balance—Kirsten Menking, Vassar College, Science Education Resource Center, On the Cutting Edge

http://cleanet.org/resources/43022.html

3. BIOLOGICAL PROCESSES DEPEND ON ENERGY FLOW THROUGH THE EARTH SYSTEM.

The third principle examines

1. How photosynthesis transforms energy from the sun into organic matter from carbon dioxide and water, serving as the foundation for energy flow through food webs

2. Food as a biofuel for organisms, releasing energy for living

3. How energy decreases through food chains, with some being stored in new chemical structures, most of it being dissipated into the environment, and the Sun adding new energy to keep the process going

4. Food webs, which are most energy efficient at the level of producers and powered by photosynthesis, become less efficient higher up in the food chain

5. Ecosystems, which are constrained by the energy and matter available to them

6. How human activities are altering the energy balance of Earth's ecosystems through agriculture, consumption, and population

The idea that biology and other life sciences are in large measure about examining energy transformation and the cycling of matter through organisms and ecosystems may surprise some, but, as John Muir observed in his 1911 book, *My First Summer in the Sierra*, "When we try to pick out anything by itself, we find it hitched to everything else in the Universe" (110). Human intellect excels at critical analysis, and science is all about deconstructing processes in order to better understand them. This principle, in which biological processes are examined through the lens of energy, is important not only for understanding the processes of Earth's biosphere but also for understanding the role of human activity as a component of the biosphere, which, in many respects, has taken on a geologic-sized life of its own.

Missed Conceptions

- *Where does food come from? Well, the grocery store, obviously.* Learning where and how meat is produced is a shocking experience for many students. Telling the story of food in and throughout our lives and in the larger food webs of the planet requires addressing numerous misconceptions or widening the view of learners, who perhaps never gave much thought to the growing of food, the amount of wasted food in many communities, or the linkages and lineage of sunlight as it is transformed through photosynthesis into carbohydrates, which in turn are transformed into other foods and fuels.

Doing the Math

How do we measure food webs and biologic processes? What are the inputs and outputs of the systems involved, whether measured in tons of grain and bushels of corn, calories ingested, or biomass generated by photosynthesis in the green-up period in the spring and summer or decaying in the fall and winter? Tracking energy, first qualitatively, then numerically, through food webs, from sunlight and photosynthesis and up and out of the food chain, fosters whole systems thinking and an awareness of where humans fit in the general scheme of things.

CLEAN Recommends

Middle and High School

> Oceanic Food Web—Office of Biological and Environmental Research of the U.S. Department of Energy Office of Science, Biological and Environmental Research Information System (BERIS)
>
> > http://cleanet.org/resources/43459.html

> The Ocean's Green Machines—NASA
>
> > http://cleanet.org/resources/43137.html

> How Cheeseburgers Impact the Oceans—Jennifer Provencher, Bamfield Marine Sciences Centre
>
> > http://cleanet.org/resources/42717.html

> How Much Energy Is on My Plate?—Lane Seeley, Seattle Pacific University; Karin Kirk, SERC, CLEAN Community Collection
>
> > http://cleanet.org/resources/44518.html

College

> Daisyworld Model—James Lovelock, Andrew Watson, and Dave Bice, Department of Geosciences, Penn State University
>
> > http://cleanet.org/resources/14136.html

> From Pond Scum to Power—Melissa Salpietra, NOVA scienceNOW
>
> > http://cleanet.org/resources/43139.html

4. VARIOUS SOURCES OF ENERGY CAN BE USED TO POWER HUMAN ACTIVITIES, AND OFTEN THIS ENERGY MUST BE TRANSFERRED FROM SOURCE TO DESTINATION.

This principle means

1. Humans use sources of energy, including fossil fuels, biomass, and renewables such as sunlight, wind, moving water, and geothermal, to transfer and transform energy for human endeavors

2. Human consumption of energy has limits and constraints, such as natural resources, policies, and socioeconomics

3. Fossil and biofuels are originally sunlight captured and transformed by photosynthesis and other processes

4. Transportation of fuels and transmission of electrical energy allows energy to be transformed and transferred over distances

5. Electricity, which is an energy carrier, not a primary source of energy, can be generated in a variety of ways

6. Energy can be stored for later use in numerous ways, including batteries, hydrogen, and thermal storage, each with their own challenges

7. Energy systems all have benefits and drawbacks, including efficiency, cost, and environmental risk

Appreciating how humans are dependent on energy from the Sun, food, and other natural resources doesn't require mastering the laws of thermodynamics. In fact, the NGSS launch into investigations of energy in our lives in kindergarten and continue through the primary grades, while delving more deeply into the complex energy equation in secondary grades.

Missed Conceptions

- *We can just get our energy from solar panels and wind turbines without much effort.* Actually, our societal dependence on fossil fuels is deeply rooted, difficult to change, and daunting. Moreover, storage of energy and the need to address peak use energy demand (certain times of day and certain times of the year) remains a challenge.

- *If we just had (fill in the blank), all of our energy problems would be solved.* Whether renewables, fuel cells, fifth generation nuclear power, or carbon capture and sequestration, all systems for generating energy for human consumption have their positives and negatives, so to speak.

- *Fossil fuels just happen to be in the ground and there's no connection to Earth's past history.* In fact, fossil fuels are very concentrated forms of buried solar energy from millions of years ago.

Doing the Math

More than perhaps any other principle with the possible exception of Principle 6, which focuses on the amount of energy used by human society, this principle lends itself to measuring energy in our lives. Establishing a baseline, whether in our individual lives, homes, schools, work environments, communities, or as a nation, is key to being able to measure progress and change. The U.S. Energy Information Administration is the official and authoritative source of information for the United States, with local and regional data sometimes more challenging to find. The International Energy Agency (IEA), "an autonomous organization [made up of 29 nations including the United States], which works to ensure reliable, affordable and clean energy," is also a source of current data and reports on world energy consumption and trends. Based in Vienna, Austria, and formed in response to the 1973–1974 oil crisis, the IEA tracks energy around the world.

CLEAN Recommends

Power Source—Steven Semken, SERC—On the Cutting Edge Collection

http://cleanet.org/resources/41901.html

U.S. Energy Production and Consumption—Environmental Literacy and Inquiry Working Group, Lehigh University

http://cleanet.org/resources/43492.html

Energy For You—ScienceNet Links AAAS

http://cleanet.org/resources/42710.html

From Grid to Home—Marie Johnson, SERC—On the Cutting Edge Collection

http://cleanet.org/resources/41900.html

Investigating Renewable Energy Data From Photovoltaic (PV) Solar Panels—Carla McAuliffe, Rita Freuder, TERC

http://cleanet.org/resources/43470.html

Natural Gas and the Marcellus Shale—Sid Halsor, SERC—On the Cutting Edge Collection

http://cleanet.org/resources/41845.html

Other Sources of Data

EPA's eGRID

Energy Information Administration

5. ENERGY DECISIONS ARE INFLUENCED BY ECONOMIC, POLITICAL, ENVIRONMENTAL, AND SOCIAL FACTORS.

These factors include the following:

1. Energy-related decisions are made on every level of society—on an individual, community, national, and international level.

2. Changing energy infrastructure is difficult because of the money, time, and technology involved, with decisions of one generation providing opportunities as well as limits on the range of possibilities open to future generations.

3. Using a systems-based approach that weighs all costs and benefits can improve energy decisions.

4. Social factors, including ethics, morality, and social norms, as well as economic factors, including subsidies, political, and environmental considerations, influence energy decisions.

Missed Conceptions

- *The problem is too big and there's nothing I can do about it as an individual.* The problems are, indeed, ginormous, to use a technical term. But we all have to start where we are, investigating and becoming aware of energy in our own lives, beginning at the beginning.

- *The problem is so huge, only by working at the top level will anything be accomplished.* Yes, the challenge does require decisions and action at the highest levels of society, but individual people should be afforded the opportunity to learn and decide for themselves and ideally support or oppose proposals and policies they don't agree with. History shows us many examples of one person making a difference. Indeed, individuals have initiated many efforts at the local and regional level.

- *The problem is caused by (fill in the blank) and we can't do anything about the problem until we stop (fill in the blank).* There are many excuses to avoid thinking about or taking action around energy waste and negative impacts from energy consumption, but learners inherently are interested in discovering their locus of control and testing where they can make a difference and where they simply can't. Blaming nations or corporations may distract from the reality that 20% of the people on the planet are responsible for about 80% of the fossil fuel emissions, and they also, in theory, have the ability to make informed energy (and climate) decisions that will reduce emissions.

- *The environment is (more/less) important than the economy.* Will transforming the global economy away from fossil fuels to a more decarbonized society cost or create more jobs? Will it be painless or easy? Should we focus more on adaptation and less on trying to move away from fossil fuels? These are questions that are better discussed in social studies and civics classes and may distract from teaching science. First things first: Let us understand the nature and scope of the challenges and, using creativity, innovation, and clear intentions, move forward in a productive, proactive way.

Doing the Math

Facts and numbers matter. Decisions should, ideally, include an analysis of costs and benefits that include environmental, social, political, and economic factors carefully weighed. But decision making is not necessarily easy or linear and can become heated and emotional. Numbers can be selectively used to buttress nearly any argument, so while numbers are vital to making informed choices and considering risks, they are only part of the equation. Other human factors, including age level, culture, current events, and communication skills, must also be brought into the mix.

CLEAN Recommends

Because of the interdisciplinary nature of this principle, CLEAN suggests opportunities to teach and learn about its concepts in political science, social studies, mathematics, economics, and/or psychology classes, where related questions can be examined through the lens of these disciplines.

Role-playing holds particular potential for exploring the societal dimension of this principle and its underlying concepts.

Car of the Future—Jeff Lockwood, NOVA Teachers
    http://cleanet.org/resources/41863.html

Great Energy Debate—National Energy Education Development (NEED)
    http://cleanet.org/resources/43021.html

Automobile Choices and Alternative Fuels—Amy Gambrill, PBS NewsHour

http://cleanet.org/resources/41858.html

Evaluating the Effects of Local Energy Resource Development—Devin Castendyk, SERC—On the Cutting Edge Collection

http://cleanet.org/resources/41912.html

Greenhouse Emissions Reduction Role—Play Exercise, K. M. Theissen, University of St. Thomas, Pedagogy in Action Collection from SERC

http://cleanet.org/resources/42716.html

6. THE AMOUNT OF ENERGY USED BY HUMAN SOCIETY DEPENDS ON MANY FACTORS.

The underlying concepts of this principle focus on how

1. Conserving energy—decreasing use of societal energy—is different from the law of the conservation of energy, which relates to the amount of energy in the universe being constant

2. Conserving energy reduces wasted or lost energy

3. Demand for energy for human activities is increasing with development and population growth

4. Earth's energy resources are limited and demand causes stress on natural processes

5. Innovation can impact the amount of energy used by society, and decreases can be beneficial

6. The amount of energy consumed can be impacted by behavior and technological design

7. Energy is embedded in products and services, and the source and amount of energy should be calculated to determine its impacts and consequences

8. Measuring energy consumed is essential to understanding costs and efficiencies of energy, whether at home, in business, or in transportation

Missed Conceptions

- *If the law says energy can't be destroyed, why should we worry about conserving energy?* While energy can't be destroyed, it can certainly be wasted or lost, which is why focusing on energy efficiency is important. While it is impossible to reach 100% efficiency, substantial strides have been made in tightening up many systems to minimize waste and find good use of by-products that otherwise would be discarded.

- *If we just focus on energy we consume directly in our lives, we'll be able to fix the problem.* A significant portion of the energy we consume is hidden away in the products and services we use. In some cases, the packaging and containers of a product may have more energy embedded in it than the product itself. An example would be a bottle of soda, where the amount of energy used to manufacture the bottle is more than the manufacture of the beverage itself or the transportation involved.

Doing the Math

While researchers have been able to aggregate data on total amounts of energy consumed and the fuel source of the energy involved, it has been difficult if not impossible to fully track the amount and type of energy (fossil fuel, renewable, etc.) of a particular product or service. But that is beginning to change. Life cycle assessment or LCA is a way of analyzing and quantifying the energy embedded in a product from the time it is created from raw materials (the cradle) to the time it is finished and discarded (the grave).

The potential for LCA to provide consumers, regulators, and businesses themselves with what Daniel Goleman (2009) has described as radical transparency is massive and game-changing but has not yet reached the level of sophistication where it can easily be used for education audiences. The complexity of tracking the energy as well as related toxic waste, water, and social footprints of products with global supply chains is enormous and daunting to piece together, but efforts such as the Sustainability Consortium involving large retailers, such as Wal-Mart and the companies that provide them with products, are beginning to make that data accessible.

Until such tools are available for students to use in their own analysis of products (which in a few years time will allow them to scan a product barcode with their phone and see the energy, water, and other footprints), it is possible to get a glimpse of the potential for this type of analysis using the U.S. Government's BEES software, which stands for Building for Environmental and Economic Sustainability, developed by the National Institute for Standards and Technology (NIST).

CLEAN Recommends

Middle and High School

> Plugged in to CO2—Lisa Gardiner, Marie Johnson, and Jonathan Hoffman, Windows to the Universe
>
> > http://cleanet.org/resources/41859.html

> Home Energy Quiz—Connecticut Energy Education
>
> > http://cleanet.org/resources/41906.html

> Are You An Energy Efficient Consumer?—Marian Koshland Science Museum of the National Academy of Sciences
>
> > http://cleanet.org/resources/41841.html

> Zero-Energy Housing—Jonathan MacNeil, Malinda Schaefer Zarske, and Denise W. Carlson, Teach Engineering by the Integrated Teaching and Learning Program
>
> > http://cleanet.org/resources/41890.html

College

> Energy Consumption Rates across the USA and the World—Glenn A. Richard, SERC—Pedagogy in Action Collection
>
> > http://cleanet.org/resources/41902.html

Gapminder: Unveiling the Beauty of Statistics for a Fact Based World View—Gapminder.org

  http://cleanet.org/resources/42499.html

7. THE QUALITY OF LIFE OF INDIVIDUALS AND SOCIETIES IS AFFECTED BY ENERGY CHOICES.

The final principle of *Energy Literacy* focuses on choices, specifically how they relate to

1. Economic security at the level of individuals and society

2. National security, especially if the nation depends on external sources of energy

3. Environmental quality, including the health and well-being of humans and other organisms

4. Reliance on fossil fuels, which are finite, as the primary source of energy for people on the planet

5. Access to energy and how it is essential to quality of life, including health, education, equality, status, and healthy environment

6. Vulnerable populations, who may benefit the most from the benefits of access to energy but be most harmed by the negative economic, social, and environmental consequences of energy choices

Making informed energy choices is on one hand simple—kindergarten students would likely agree that everyone should have access to be able to meet essential energy needs—and yet also complex and challenging. Ideally, learners will by graduation have an understanding not only of the essential physical science of energy but also of the broad environmental, societal, and economic context and implications involved.

Missed Conceptions

- *If we allow people in poverty to have our lifestyles, it will be game over for the climate system.* Meeting essential energy needs of the billions of people on the planet living without electricity and/or clean water is challenging but doesn't require replicating the energy infrastructure that much of the developed world has come to be dependent on. Solar energy is particularly well suited for parts of Asia and Africa where population growth and poverty is the most extreme, although energy storage is still a technological challenge.

- *We can't afford to help the rest of the world with their energy challenges since we have enough problems of our own.* While getting our own house in order is clearly a priority, many people feel we have a responsibility to help others who are vulnerable or less fortunate. Finding the most effective way of doing so without having good intentions go awry is especially challenging.

Doing the Math

What are ways of measuring economic and national security? What metrics are valid for weighing environmental impacts, ecosystems benefits, and societal costs of energy and climate? These are questions that top economists and experts wrestle with, but they also offer learners real-world problems that can challenge their critical thinking, numeracy, and engineering problem-solving abilities. As the costs of renewable energy drop and fossil fuels rise, the cost-benefit and return on investment equations change, sometimes quickly. Helping students access such data and come to their own conclusions is important toward building climate and energy literacy.

CLEAN Recommends

Middle School Through College

> Samoa Under Threat—Andrea Torrice, Bullfrog Films; Teachers' Domain
>> http://cleanet.org/resources/42767.html

> Equity and Climate—Kate Drake, Kelsey Poole, Earth Day Network
>> http://cleanet.org/resources/41873.html

> Energy and the Poor—Black Carbon in Developing Nations—Anne Hall, Environmental Studies at Emory University, from the On the Cutting Edge Collection
>> http://cleanet.org/resources/42706.html

> World Climate: A Computer-Simulation-Based Role-Playing Exercise—Fiddaman, John Sterman, copyrighted by Schlumberger Ltd.
>> http://cleanet.org/resources/43001.html

In conclusion, teaching and learning about energy takes, well, energy and creativity. And knowing the essentials of energy is, yes, essential to being able to convey how it works and why it is vital to our understanding of the universe in general and our future in particular.

Long sequestered away in physics courses rather than used as the dynamic vehicle for bringing all manner of topics alive, it is high time to bring energy out of the dreary textbooks of yore and into our lives. Energy can spark our imagination about what is possible even as it overwhelms us with the daunting scale and scope of our collective energy needs. Confronting the energy crisis and climate change requires substantially increasing our individual and collective energy literacy.

ADDITIONAL RESOURCES

Bakst, B. (2013, September 27). Solar garden: Model T of renewable energy? *Christian Science Monitor*. Retrieved from http://www.csmonitor.com/Environment/Latest-News-Wires/2013/0927/Solar-garden-Model-T-of-renewable-energy

Belsie, L. (n.d.) Think you know energy? Take our quiz. *Christian Science Monitor*. Retrieved from http://www.csmonitor.com/Environment/2013/0124/Think-you-know-energy-Take-our-quiz/consumption-rise

Hill, A. (2013). Does a cell phone use as much electricity as a refrigerator? *Marketplace*. Retrieved from http://www.marketplace.org/topics/sustainability/no-your-phone-doesnt-use-much-electricity-refrigerator

Unger, D. J. (n.d.). U.S. energy in five maps (infographics). *Christian Science Monitor*. Retrieved from http://www.csmonitor.com/Environment/2013/0530/US-energy-in-five-maps-infographics/1.-Oil-gas-and-coal

CHAPTER 6

Programs That Work

This fledgling learning revolution with climate and energy literacy has been set in motion by a convergence of factors: motivated learners and educators, support and inspiration from leaders and institutions, seed funding, and entrepreneurial creativity. From preK through graduate school and beyond and in communities, government agencies, and businesses large and small, climate- and energy-related topics and skills are being deployed, from the bottom-up, top-down, inside out.

Given the urgency and scope of the challenges and the general lack of coordination and large-scale funding, these efforts are often bootstrapped and have not yet been sufficient to spark critical mass transformation in society that many say is needed for us to address the scope of the daunting challenges we face. But they offer glimpses of what is possible and in some cases provide powerful exemplars of what education can and should look like in the 21st century.

A burst of federal grant funding for climate change education in 2009 jump started curriculum development and teacher professional development programs, but most of those programs have languished once their funding was gone. Among the initial projects that were funded by the National Science Foundation were CLEAN (http://cleanet.org), CAMEL (http://www.camelclimatechange.org), resources primarily for undergraduate students, and Earth: The Operators' Manual (ETOM for short, http://earththeoperators manual.com), developed by Geoffrey Haines Stiles, producer of the original *Cosmos* television series with Carl Sagan. ETOM was first a PBS NOVA television program featuring Penn State geologist and expert Richard Alley. The original program grew into a robust website, additional episodes, teaching tips, and a companion book, which several universities are now using as a text for incoming freshmen to read in order to help them get up to speed on climate and energy basics. As a starting place and one-stop-shop for introductory materials that emphasize a can-do attitude, ETOM is hard to beat.

NASA, which has its engaging climate website (http://climate.nasa.gov) developed by the Jet Propulsion Laboratory, has also funded over 70 climate education projects since 2008 and currently focuses its programs on increasing access of underrepresented minority groups to science careers and educational opportunities through its NICE (NASA Innovations in Climate Education) Program: https://nice.larc.nasa.gov.

Other federal agencies have also contributed to climate, including NOAA's Climate.gov teaching section (http://www.climate.gov/teaching) and the U.S. Forest Service's

Climate Change Live distance learning adventure (http://climatechangelive.org). Many of the resources in these programs have been incorporated into the CLEAN Collection (http://cleanet.org), and most have been evaluated to assess their scientific and pedagogical integrity.

One ambitious project that has been years in the making is the National Science Foundation's Climate Change Education Partnership program, which is made up of six projects from across the nation that are coordinated through an alliance office at the University of Rhode Island: http://ccepalliance.org.

The six projects were part of a competitive process that began with over a hundred projects vying for funding to develop strategic plans. Through a panel review process, 15 were selected for the first phase, and this number was then reduced in the next round to the following six, which were chosen to deploy their strategic plans.

- **Climate Education Partners** (CEP, http://www.sandiego.edu/climate), led by the University of San Diego, aims to increase the understanding among key influential leaders in San Diego (such as business leaders and real estate brokers) about the causes and consequences of climate change, to evaluate the most effective methods to convey information about the impacts on the region's economy, weather, natural resources, and air and water quality, and to help replicate their model on a national scale, bringing together scientists, educators, and community leaders at local levels.

- **Climate and Urban Systems Partnership** (CUSP, http://www.cuspproject.org) is headed by the Franklin Institute in Philadelphia with partners in New York City, Pittsburgh, and Washington, DC. It describes itself as a "network of informal educators, climate scientists, learning scientists, and local community organizations across four cities dedicated to improving local understanding of and engagement with climate change science." Their goal is to: "Engage city (or **urban**) audiences in community-wide (or **systems**-level) issues related to **climate** change via local **partnerships**."

- **MADE CLEAR**, the Maryland-Delaware Climate Change Education, Assessment and Research program (http://www.madeclear.org), is a collaboration between the University System of Maryland (USM) and the University of Delaware (UD) to implement, evaluate, inspire, and support climate change education in the two states, which are both particularly vulnerable to sea level rise. K-12 education and professional development for teachers is a primary focus. Maryland Public Television is a key partner, and both states have adopted NGSS. Each state's department of education is involved in this program.

- **National Network for Ocean and Climate Change Interpretation** (NNOCCI, http://www.nnocci.org) is made up of some of the biggest names in informal science education in the nation: the New England Aquarium, the Association for Zoos and Aquariums, the National Aquarium at Baltimore, the New Knowledge Organization, the Frameworks Institute, Penn State University, Woods Hole Oceanographic Institution, the Center of Science and Industry, and the Monterey Bay Aquarium. Through year-long study circles with scientists, the NNOCCI is working toward developing a national network of informal science education interpreters who will use research-based communication strategies to promote their mission: "Train enough voices in proven communication techniques to change the national discourse around climate change to be productive, creative and solutions focused."

- **Pacific Islands Climate Education Partnership** (PCEP: http://pcep.wested.org) focuses on U.S.-affiliated islands in the Pacific, including Hawai'i, American Samoa, Guam, the Northern Marianna Islands, the Marshall Islands, Palau, and Micronesia, with the goal of empowering the students and citizens of the region by implementing a Climate Education Framework. By infusing indigenous knowledge, the project will help meld cutting-edge science with local wisdom relating to adaptation strategies, climate, and environmental stewardship.

- **Polar Partnerships** (Polar Learning and Responding Climate Change Education Partnership, http://thepolarhub.org) is a diverse partnership headed at Barnard College, Columbia University with a focus on climate change in the polar regions through the use of a wide array of games and activities, including the following:

  o EcoChains: Arctic Crisis, a card game about marine ecosystems

  o SMARTIC (Strategic Management of Resources in Times of Change), a classroom activity on marine spatial planning

  o AMNH's Seminars on Science Climate Change course, professional development for teachers online

  o Polar Explorer, a data visualization app

  o Polar Voices, a mystery and adventure radio series

These diverse Climate Change Education Partnerships (and some of the programs that were developed but not funded) have the potential to be expanded and transferred to other communities—and potentially other nations—that are seeking high quality (ideally free) resources and effective strategies to inform and engage communities.

There are many other inspiring individuals and programs. One of the individuals making waves is mother, educator, and scientist Dr. Juliette Rooney-Varga, Director of the UMass Lowell Climate Change Initiative. Deeply concerned about human impact on the climate system and the need to make better informed decisions to prepare ourselves for changes well under way, Juliette works with top scientists at MIT to develop interactive online simulations that students around the world can use to foster a systems approach to thinking about climate and energy. She works with undergraduate students, having them learn the essentials of climate and energy and then making videos that are a form of assessment on what they have learned. She is also involved in her son's school and with her local community, Somerville, Massachusetts, which adopted a climate action plan in 2001.

Not everyone has the drive or commitment of Juliette to push for society to be climate smart and energy wise, but many other educators, parents, and teachers are doing what they can to convey key scientific content, systems dynamics, and systems thinking in order to promote the understanding and development of solutions to these cross-disciplinary domains. They are using citizen science and interactive tools to more fully engage learners, deploying immersive games and interactive strategies for skill building, tapping the potential for service learning to gain real-world problem-solving experience, and building diverse communications skills—from basic oral and written to sophisticated, state-of-the-art presentations.

Greening of Education

There are a wide array of green school programs, most of which share the word "green" but are otherwise not connected or coordinated. Among the programs that are beginning to lay the groundwork for transforming education are a spectrum of green projects:

- Green Ribbon Schools, an awards program of the U.S. Department of Education that recognizes schools and districts that are exemplary in reducing environmental impact and costs, improving the health and wellness of students and staff, and providing effective environmental and sustainability education; not to be confused with the Texas Green Ribbon Schools program that promotes health and sustainability and is not connected to the national program

- Green Schools National Network, which collaborates with green school programs in many states and hosts a large annual conference where green school programs rendezvous

- Green Schools Alliance, which aims "to connect and empower K–12 schools worldwide to lead the transformation to global environmental sustainability"

- The relatively small, Massachusetts-based nonprofit Green Schools, whose mission is to create greener and healthier students and schools through environmental education, innovation, leadership, and action

These programs are, in some respects, a legacy of the environmental and the Greening of America movement (Reich, 1970) of the late 1960s and early 1970s, when President Nixon established the Environmental Protection Agency and signed the Clean Water Act and the Occupational Safety and Health Act (OSHA) was passed by Congress. While these and related environmental education efforts are important and can help foster environmental awareness and conservation ethics, they do not necessarily include science or climate and energy literacy as explicit elements of their programs.

For Green School or other environmental education efforts to support climate and energy literacy, they need to go beyond recycling, outdoor, or school garden programs, though those can be an important start. They may need to overcome resistance to teaching and learning about science, taking full advantage of teachable moments to demystify the science. As Allison Anderson of the Brookings Institute writes in her 2010 white paper, *Combating Climate Change Through Quality Education:*

> There is a clear education agenda in climate change adaptation and mitigation strategies, which require learning new knowledge and skills and changing behaviors in order to reduce the vulnerabilities and manage the risks of climate change.

Programs that work and prepare learners for the challenges ahead share many of these strategies:

- Focusing on solid science understanding through learner-centric inquiry and engagement

- Balancing the realities of science and society with effective solutions and responses

- Emphasizing systems dynamics and thinking in a real-world context

- Weaving science and social studies content together with other disciplines, including mathematics, history, art, service learning, communications, and presentation skills

- Providing opportunities for learners to teach others, especially peers

Driven individuals initiate many programs:

- Teachers who talk with other teachers and their administrators to discuss creative ways of infusing climate, energy, and related topics throughout the curriculum

- Learners who—through their own initiative and/or mentoring and empowerment by others—take on leadership roles within their own schools and communities

- Administrators and officials who give the green light of support and perhaps seed funding to get things rolling

- Parents who help instigate the important, sometimes difficult conversations within schools and districts, to make sure these topics are taught and taught well

ACE Is the Space

One program that is devoted entirely to fostering climate and energy awareness and literacy is the Alliance for Climate Education (ACE). Based in Oakland, California, but with educators in many large metropolitan centers across the United States, ACE has reached over one and a half million students in more than 2000 schools with their engaging, multimedia assemblies for high school students that last about one class period. While short by traditional science education standards—usually less than an hour long—the assemblies have substantial impact on the understanding of those attending.

The people on the front lines of ACE's program are known as the educators, but they do much more than teach about climate and energy. The ACE assemblies are designed to be accessible, inspiring, and a little flashy, and the educators themselves are typically young, hip, and very personable. They handle logistics; hand out materials; act as a hybrid between a lecturer, performer, comedian, and substitute teacher; and then help mentor students who, inspired by the presentation, want to start an environmental club or attend a youth leadership program.

One ACE educator working out of the Oakland headquarters is AshEl Eldridge. While usually AshEl does assemblies in front of large groups of students in a high school auditorium, occasionally he presents to classes. One of those classes was Mr. Duffy's AP chemistry class at Concord High School in Concord, California, which I visited in the fall of 2013. The high school, which features a student parking area shaded by solar panels, serves about 1,500 students in the largely suburban Mount Diablo Unified School District, which now gets over 90% of its energy from solar power.

Clearly comfortable bantering with youth, AshEl was articulate and well informed about climate science. He delivered his presentation using exciting multimedia elements, which walked students through the essential science of climate change, the greenhouse effect, and how human activities, including the U.S. propensity for living large, contribute to the altered climate. He touched on the right of young people to know about climate change and suggested they check out ACE's Right to Know campaign (http://change.acespace.org), which declares the following:

1. WE HAVE THE RIGHT TO KNOW that climate change is happening, and that its impacts will affect us all within our lifetimes and beyond.

2. WE HAVE THE RIGHT TO KNOW that solutions to this challenge exist right now.

3. WE HAVE THE RIGHT TO ACT by educating our peers and family members about climate change impacts and solutions to spark a real national conversation.

The concept of having a right to know about potential hazards is a deeply held value that goes back to the Pure Food and Drug Act of 1906 (Swann, 2009), which recognized that Americans have the right to know what is in their food and drugs. Other examples include the Federal Trade Commission's monitoring of truth in advertising and the numerous environmental and health statutes designed to provide people with access to information about hazards in their communities, in the environment, and, through the Occupational Safety and Health Act (OSHA) of 1970, in their workplaces. The Right to Know campaign rests on the belief that individuals are entitled to a basic education that will prepare individuals and thereby society for the challenges and opportunities of the future.

The value of ACE is multifold. Research that ACE has conducted with experts at Stanford and Yale has found that ACE's blend of entertainment and education changes young people's beliefs, level of involvement, and behavior in a positive way, encouraging them to be more climate science literate and planting seeds that will empower them to become more engaged in school programs and to take on new challenges. Preliminary research findings suggest that this hybrid blend of education and entertainment appears to have a positive impact on beliefs, involvement, and behavior, but additional research is needed to confirm the viability of this approach. There's only so much that will stick in students' minds from a one-class period assembly, but many science teachers appreciate ACE's edutainment approach because it makes the science more accessible to students during normal class periods. Science educators like Mr. Duffy sponsor ACE assemblies on a regular basis because they motivate and provide a fresh perspective and context for the science the students are learning in science class.

ACE and programs like it are also valuable in how they inspire students beyond the classroom, giving them the motivation and tools to lead climate and energy initiatives in their schools and then become leaders in college and in their communities. For example, the Edward C. Reed High School in Reno, Nevada, had no active environmental clubs or initiatives in 2010, but after seeing the ACE Assembly, a group of motivated students, supported by two dynamic science teachers, formed an action team. They entered the GREENevada Student Sustainability Summit, researching and writing a proposal for how they'd like to green their school. Focusing on wasted energy and water in bathrooms—lights were on 16 hours a day whether or not the rooms were occupied, and faucets were constantly leaking—they calculated potential savings in gallons of water, kWh of electricity, pounds of carbon dioxide saved, and dollars not spent by installing low-flow toilets along with new faucets and lights with auto-sensors. Well prepared by their research, they won first prize at the summit—a total of $12,000 to upgrade their school's bathrooms. Impressed by their initiative, the school district chose Reed High to be the first facility to receive solar panels. With the bathroom renovations complete and a reminder of what a student initiative can accomplish, the Reed High Eco Warriors Club is going strong, looking for new opportunities to build on their success.

ACE worked with 15 students at schools in California to develop a series of impact videos showing the need to address the poor conditions of school facilities, especially as they relate to water and energy. The project successively sought funding from agencies "for improving energy efficiency and creating clean energy jobs" through California Proposition 39 funds. Ideally, these energy upgrades of schools will not only save money on energy but also improve the health of students and staff as well as the educational experience.

Chloe Maxmin is an example of a student who believes in action rooted in understanding. Growing up on a farm in rural Maine, Chloe organized an ACE Assembly at Lincoln Academy in Damariscotta, Maine, for 600 fellow students. This led to her founding the high school's Climate Action Club, whose accomplishments include keeping 700,000 plastic bags from local landfills and pushing for the installation of solar panels at the school. She started First Here, Then Everywhere, a network for young environmentalists. After moving on to Harvard, she has led the 350.org divestment campaign on campus, which calls for the university to divest its financial holdings in fossil fuel corporations. Her story has been featured in *Rolling Stone* as a "Green Hero," on Boston's WGBH News, and on the ACE blog. Having a solid understanding of the essential science has helped Chloe make her own informed choices and priorities.

SUMMER SCHOLARS

It's 9 a.m. in early June shortly after the school year has concluded, but the cafeteria of a school outside of Lafayette, Colorado, is filled with 80 excited middle school students gathered for the Dawson Summer Institute. Some of the taller students tower over Lisa Michaels, the petite and driven leader of the program, as she opens the three-week initiative—Climate: Vital Signs of the Planet—a program for high-achieving students. Michaels sets clear and high expectations in terms of behavior for the students she calls scholars. Representing a creative hybrid of public and private school collaboration, the program is funded by the Dawson Foundation, which also supports the private Dawson School.

In past years, the institute's initiatives have focused on different themes, including food and energy. This year, by popular demand, the theme is climate. Drawing on experts to help with the scientific content, videographers to teach production skills, and science educators to help facilitate the process, the scholars are immersed in the science and related societal issues related to the climate. Along the way they are learning collaborative and communication skills, which they will use when presenting their findings. Final presentations in the past have included campaign posters, multimedia presentations, and even songbooks encouraging hand washing. The scholars were divided into four teams and produced videos, presentations, and posters to share with the community.

The Climate by Design team explored

- eco-friendly, energy-efficient building construction, including learning how a home can be cooled and heated by various passive systems;
- alternative transportation, such as maglev trains and hydrogen fueled buses, looking at statistics on transportation-related carbon emissions;
- electrolysis, as demonstrated by splitting hydrogen from water; and
- making of biofuels.

The Pulse of the Planet group focused on ecology and climate and developed

- a Jeopardy!-style trivia game;
- animation on the carbon cycle; and
- visual representations depicting increasing/decreasing populations of species affected by climate change.

The Leave No Trace? team focused on conveying to the public issues relating to

- wildfires;
- black carbon from cook stoves;
- awareness of climate change using the model of the Six Americas segments of the American public that interact with the issue of global warming in their own unique ways; and
- adaptation and mitigation programs in different regions in the United States.

The fourth and final team, Power to the People, developed an array of different materials, including

- model UN Project Climate, producing curriculum, posters, and a simulation game;
- history of the electric grid;
- teen TED Talk about Climate Science; and
- wave energy, for which they produced three posters and a hand-made wave machine to help convey their findings.

By the end of the three weeks, the scholars gained a solid understanding of the complexities of climate change and related challenges; learned new technical, collaboration, and presentation skills; and made new friendships during the institute.

Maximizing Teachable Moments/
Schools as Living Laboratories

When the students in Aaron Sebens's fourth-grade class at Central Park School for Children—a public charter school in Durham, North Carolina—were learning about solar energy, someone had the idea that the class should have their own solar panels. So they set out to raise $500 through a Kickstarter campaign, but they ended up raising $5,000 from around the world. Schools elsewhere, often inspired by one student, parent, or teacher, have embarked on similar projects, raising money for solar panels and dashboards that track energy consumption through mini-grants and donations, providing the school with ongoing teachable moments.

Casey Middle School in the Boulder Valley School District, which was renovated in 2010, provides live data of energy consumption and a virtual tour of how their school has been transformed and updated for the 21st century. Originally built in the 1920s, in 2006 voters approved a $296 million bond for capital improvements. The result: Casey Middle School was one of two LEED Platinum schools in the nation when the renovation was completed. The district has also hired a sustainability coordinator who, among other things, interacts with the energy and climate efforts of the city and county of Boulder and the University of Colorado at Boulder.

In California, Prop 39—the California Clean Energy Jobs Act passed by voters in 2012—will provide an estimated $2.75 billion over five years to the state's public schools for energy efficiency and alternative energy projects. The money, gained by closing a tax loophole on out-of-state corporations, is not dedicated to teaching or curriculum, but as school infrastructure is updated, the ripple effect throughout the learning community that the building serves will be substantial.

While the money from Prop 39 is a drop in the bucket of the $70 billion state education budget (California Department of Education, 2013) in 2012 through 2013, every drop helps. A UC Berkeley study found that $117 billion is needed over the next decade to deal with the facility infrastructure of schools in California. Seventy percent of the state's school buildings are over 25 years old and 30% are more than 50 years old, and like many districts in the nation, a large percentage of districts in California have not been able to raise a bond to improve their facilities in several decades, leaving infrastructure run down and decayed. The Center for Green Schools in their 2013 State of the Schools report estimates that $271 billion is needed to bring K–12 school buildings up to working order and comply with laws, and another $542 billion is required to invest in modernizing schools to meet current education, safety, and health standards.

The win-wins from upgrading run-down schools, making them more energy efficient and ideally adding renewable energy, are substantial. Energy savings provide cost savings; one school that upgraded its facility has already restored its cancelled music program through the money saved (McCaffrey, 2013). Cost savings aren't the only advantages: The physical quality of education benefits too. Improving the HVAC systems in schools enhances air quality, which in turn leads to reduced asthma—a leading cause of absenteeism. The data are clear: Better ventilation and lighting can improve students' academic performance.

And although curriculum is not an explicit part of California's Proposition 39, the teachable moments that Prop 39 can help tease out and catalyze are everywhere. Districts, like Sac City—one of the largest districts in the state, covering much of Sacramento—already have related programs like Project Green and Edible Sac High, which have shown potential for engaging students in innovative ways. Improving the energy systems of school facilities has also been a catalyst for students who engaged in a competition between twelve campuses in the district to audit waste and energy and then recommend improvements, with winning teams then having their proposals funded.

Gardens at schools and energy audits, as engaging as they may be, are unnecessarily limited if climate isn't brought into the mix. Schools in colder climates, where the growing season begins around the time school gets out in the spring and lasts until school begins again in the fall, may have to limit their gardening to a greenhouse or grow hole (a hole dug in a sun-facing slope that is covered with window materials that allows the growing season to be extended). In other regions, nearly year-round growing seasons can allow students to grow and cook their own food as part of their learning experience. Like gardening, energy audits offer an opportunity to examine seasonal changes and climatic dynamics, providing a context for inquiry questions, such as what are the growing opportunities this time of year? And what are the current and projected heating or cooling needs of the school or the wider community at different times of year?

GREENING THE IVORY TOWER

Where climate, energy, and sustainability literacy are actually starting to mature is in higher education. That is not to say most students graduating from college today have ever learned the basics of climate change by any means. But many schools are plunging into climate and energy literacy head first, including

- Community colleges like Butte Community College, which has become grid positive, and Bunker Hill Community College, which has enacted a 24-hour schedule to maximize the use of their facility and offer their students more classes at times literally throughout the day

- Liberal arts colleges like Furman and Unity that are integrating sustainability throughout the academy

- Large research universities like Arizona State, that are taking full advantage of bountiful solar energy with a variety of initiatives and the largest solar energy array of any campus in the United States, or the University of Colorado at Boulder, which has begun to weave sustainability throughout the curriculum and offers certificate programs in sustainable practices

Many of these colleges are part of the American College and University College Presidents' Climate Commitment (ACUPCC) or, as most call it, the Presidents' Climate Commitment. The presidents or chancellors of roughly 700 academic institutions have signed the commitment, which is a statement affirming the seriousness of climate change, the imperative of addressing its implications, and the responsibility of higher education to step up to the challenge. By signing it, these leaders are agreeing to develop and deploy carbon neutrality plans as well as offer related courses and other educational experiences to all students.

Some colleges and universities haven't, for a variety of reasons, signed the commitment. One college president felt committing future presidents of his institution to cutting carbon to zero was unrealistic or possibly unfair. But many have found that the combination of leadership from the top, the pushing and prodding from students at the grassroots, and the support of faculty and especially facilities staff can be a winning formula for transforming the academic institution from the inside out. For example, Dickinson College stresses sustainability throughout the school, which is meant to be a living laboratory, "defining real problems and real solutions." Related programs at the Carlisle, Pennsylvania, school include a Center for Sustainability Education and a Center for Global Study and Sustainability.

In the 2012 report summarizing the first five years of success, the Presidents' Climate Commitment found that 298 schools produced some 295,000,000 kWh of renewable energy annually, and 113 schools acquired over $209 million of external funding for related projects. In addition, 115 schools require all students to have sustainability as a learning objective, 82 offer professional development to faculty for sustainability education, and 68 incorporated sustainability learning outcomes into their institutional general education requirements. Collectively, the ACUPCC community includes 11,626 faculty members who are engaged in sustainability research.

Many of these same colleges also are active with the Association for the Advancement of Sustainability in Higher Education, or AASHE, whose STARS tool—Sustainability Tracking, Assessment and Rating System—helps measure progress with metrics that can then be tracked throughout the wider community.

Another effort that is gaining some traction on college campuses is the movement to promote divestment of fossil fuel investments, which is being promoted by 350.org and a spin-off organization called http://gofossilfree.org.

BALANCING ACT

There are dangers to overselling or oversimplifying solutions, and there will be critics who will accuse these efforts to jumpstart climate and energy literacy as covert attempts to achieve political ends or to brainwash impressionable young people to become eco-activists. But schools, as the NGSS demonstrate, are appropriate places to get students thinking in a scientific way and to consider the implications of the science they learn. Educators not only provide students with differing points of view on an issue, even when one side is more compelling than the other, but most importantly, they must also teach them the skills to evaluate claims and analyze data, enabling students to make informed decisions. Therefore, it is vital that educators find a balance, informing, engaging, and inspiring learners without glossing over the complexity or unintended consequences of either the implications of the science or the proposed solutions.

Ideally all learners will acquire the basics of climate and energy, be able to make informed decisions, and engage in discussion and honest debate about the range of technological, political, ethical, and practical responses to climate and energy challenges. But we live in a less than ideal world, and realistically, progress toward improved literacy will likely come in fits and starts. At Peak to Peak, a K–12 Charter School in Lafayette, Colorado, the idea of focusing on climate change excited science coordinator Adam DiGiacomo when he first began talking to faculty about it several years ago. One of the math teachers was already using climate and energy to teach math, but a physics teacher was skeptical about how carbon dioxide measurements were being made and worried that the science was overblown. It literally took several years of long conversations, developing trust among faculty and talking through the nuances of the science to fully integrate climate, energy, sustainability, and conservation into the school's curriculum. But today, Peak to Peak, rated as one of the top public schools in the nation, graduates students that have attained a solid literacy around climate, energy, and related issues.

The goal of the Next Generation Science Standards is similar: that all students gain an understanding of the essential science and related technology and engineering practices, enriched by language arts, mathematics, and other domains of learning and creativity, which will prepare them for the 21st century. This knowledge will foster the links between society, science, technology, and engineering and clarifying related problems and design solutions.

In Chapter 2 we focused on how it is essential to make climate and energy relevant, local, human, pervasive, and hopeful. We can add a few more strategies that can help make these challenging topics meaningful and accessible to learners:

- Learn by teaching. There is no better way to demonstrate mastery of a topic than by teaching it. And peer-to-peer learning is often how learning occurs, whether it is students in a study group helping each other understand the nuances of a particular concept, or workers helping one another perform more effectively.

- Learning to learn—being open minded and willing to admit when you need help understanding—is, in all aspects of life and especially in topics like climate and energy where the complexity is enormous and new insights and developments occur almost daily, essential.

- Flipped classrooms (Tucker, 2012) or flip teaching, whereby learners watch lectures on videos on their own and then in classroom there is time for collaborative learning and personal attention, are also contributing to this new learning paradigm.

- Professional Learning Communities (PLC) or similar groups of teachers can support and mentor each other with the shared goal of improving student performance and maximizing their learning experience.

Ultimately, everyone in the community, everyone in the nation, is a stakeholder and has a vested interest in making schools climate safe, not only from direct climate change but also from all types of natural hazards, certainly including storms and earthquakes. We all have an interest in transforming schools into living laboratories as well, investing in the future of our children and our collective future by making sure we provide children with the learning environment, the knowledge, and the analytic skills they need to not just survive but truly thrive in the years and decades to come. In doing so, we can also help them deal with the concerns and their own reactions to those concerns, including denial and even despair, as we'll discuss in Chapter 7.

Finally, there are a number of high-quality resources and associated communities that educators can benefit from. In addition to CLEAN, the California-focused Environment and Education Initiative offers over 80 curriculum units, each meant to be a two-week supplement to enrich science or social studies classes. Other states have expressed interest in developing their own place-based modules using a similar approach (http://www.californiaeei.org).

Also of note is the Lawrence Hall of Science's Global Systems Science curriculum for high school, which covers a wide range of global change topics including climate change (http://www.globalsystemsscience.org).

Clearly, there's no shortage of creative ideas and strategies, but there are certainly, as we will examine in the next chapter, some underlying challenges that have been obstacles to wider transformation of schools into living laboratories for the 21st century.

ADDITIONAL RESOURCES

Cohen, T., & Lovell, B. (n.d.). *Campus as a living lab: Using the built environment to revitalize college education.* Retrieved from SEED, American Association of Community Colleges, and Center for Green Schools http://theseedcenter.org/Resources/SEED-Resources/SEED-Toolkits/Campus-as-a-Living-Lab

Environmental Protection Agency. (2009). *Schools as living laboratories, energy efficiency programs in K–12 schools: A guide to developing and implementing greenhouse gas reduction programs.* Retrieved from http://www.epa.gov/statelocalclimate/documents/pdf/k-12_guide.pdf

This is a must read for administrators or educators interested in making their schools more efficient and less wasteful.

Environmental Protection Agency. (2012). *Energy use in K–12 schools.* Retrieved from the EPA's Energy Star program http://www.energystar.gov/ia/business/downloads/datatrends/DataTrends_Schools_20121006.pdf

The booklet highlights data trends for the 51,500 K–12 school properties in the nation that have a collective footprint of 5.4 billion square feet. Becoming tuned into the heating and cooling degree days and related energy demand over the course of the school year, traditionally the domain of the facilities managers of schools, can be an educational opportunity for all concerned: staff, faculty, and certainly students.

CHAPTER 7

Countering Skepticism, Denial, and Despair

The alarming prospect of not only a hotter world, higher sea levels, more intense storm events, and flooding, but also increased disease, fresh water and food shortages, supply chain disruptions, and other national and global security stresses may, understandably, trigger a variety of reactions, from denial and despair to fear and sadness.

For some, there's an understandable tendency to look the other way and focus on more immediate concerns. One natural response to being confronted with climate change bad news is to doubt the science or question the scientists. Can it really be as bad as all of that? Are the scientists mistaken? Could they be missing something? Such skepticism is natural and important. Educators often struggle to encourage healthy skepticism and critical thinking among their learners, so when introducing climate and energy topics, finding out what naïve ideas, misconceptions, or doubts they have is an important starting point.

Knowing that skepticism is the lifeblood of science, John Cook, an Australian scientist long interested in climate change, started his website Skeptical Science in 2007 as a response to comments by U.S. politicians that climate change was a hoax. Over the years, Cook (2014) and his team have assembled a collection of over 170 arguments against the hoax charge. Their site (http://www.skepticalscience.com) is used by educators and even scientists not familiar with all the current research and is especially helpful in being able to counter common statements that arise in discussing climate and energy issues.

Taking a layered approach by first presenting the myth, then providing a basic as well as more intermediate understanding of the current science countering the myth, the Skeptical Science team analyzed the merit of the arguments in detail. Many of the myths originate from legitimate questions asked by skeptical novices and experts alike who are trying to wrap their minds around and inquire about a topic that is immense in scope and complexity. Responses to twelve of the more common arguments against human-induced climate change that Cook's team deconstruct and respond to on their website are captured in the table below.

	SKEPTIC ARGUMENT	VS.	WHAT THE SCIENCE SAYS
1	Climate has changed before.		Climate reacts to whatever forces it to change at the time; humans are now the dominant force.
2	It's the sun.		In the last 35 years of global warming, sun and climate have been going in opposite directions.
3	It's not bad.		Negative impacts of global warming on agriculture, health, and environment far outweigh any positives.
4	There is no consensus.		Ninety-seven percent of climate experts agree humans are causing global warming.
5	It's cooling.		The last decade, 2000-2009, was the hottest on record.
6	Models are unreliable.		Models have successfully reproduced temperatures since 1900 globally, by land, in the air, and in the ocean.
7	Temperature record is unreliable.		The warming trend is the same in rural and urban areas, measured by thermometers and satellites.
8	Animals and plants can adapt.		Global warming will cause mass extinctions of species that cannot adapt on short-time scales.
9	It hasn't warmed since 1998.		For global records, 2010 is the hottest year on record, tied with 2005.
10	Antarctica is gaining ice.		Satellites measure Antarctica losing land ice at an accelerating rate.
11	An Ice Age was predicted in the '70s.		The vast majority of climate papers in the 1970s predicted warming.
12	Carbon dioxide lags temperature.		Carbon dixoide didn't initiate warming from past ice ages, but it did amplify the warming.

NOTE: For more, visit: http://www.skepticalscience.com/argument.php

Cook and his team relied on peer-reviewed literature in their rebuttals of the arguments against the science. If there are new insights or findings from scientific research in the literature, Cook and his team update their commentary.

Many educators and others interested in climate change have found Skeptical Science and a similar, somewhat more technically oriented website called Real Climate (http://www.realclimate.org), started in 2004 by nine climate scientists, to be invaluable. The sites help rebut arguments purporting climate change to be a hoax or that human activities couldn't possibly be the cause of the heating of the atmosphere and ocean. They also help users in their own skeptical inquiries into how scientists know what they know.

Cook's original theory of change was based on the notion that countering confusion or misinformation, whether legitimate skepticism or manufactured doubt motivated by a political or cultural bias, could be achieved by a logical, reasoned approach: Explain the myth, then analyze it in light of the scientific literature, and as a result, light bulbs of understanding will go off. Mission accomplished. As experienced teachers or parents will attest: If only it were so easy! Climate and energy issues are not only complex but also replete with psychological, social, cultural, political, global, and individual predilections. As we will see in a moment, Cook has rethought the challenges of countering climate confusion, but first, let's look at an overview of the spectrum of climate confusion.

DENIAL, DOUBT, AND MORE

In 2008 a series of studies was initiated that have become known as the Six Americas studies. Conducted by the Yale Project on Climate Change and the George Mason University Center for Climate Change Communication, these surveys have measured the beliefs, attitudes, values, and more of American adults regarding climate change. The study's report identifies Global Warming's Six Americas: six unique audiences within the American public that each responds to the issue in their own distinct way. The reports define the audience segments as the Alarmed (16% as of the January 2014 report of a survey conducted in November 2013), who are fully convinced of the reality and seriousness of climate change and are already taking action to address it. The Concerned (27%)—the largest of the six Americas—are also convinced that global warming is happening and is a serious problem, but they have not yet engaged the issue personally. Three other Americas—the Cautious (23%), the Disengaged (5%), and the Doubtful (12%)—represent different stages of understanding and acceptance of the problem, and none are actively involved. The final America—the Dismissive (15%)—are very sure it is not happening and are actively involved as opponents of a national effort to reduce greenhouse gas emissions.

Significantly, three in four of the Alarmed "often" or "occasionally" talk with family and friends about the topic. By contrast, only one in four of the Concerned do so, and 90% of the other groups indicate they discuss the subject only "rarely" or "never" (Leiserowitz, Maibach, Roser-Renouf et al., 2014). Improved literacy will inevitably increase the frequency and depth of such conversations, not only in classrooms but also around the kitchen table and the conference table.

The Six Americas reports are treasure-troves of information about people's attitudes and opinions, and we will examine how they can be applied in learning environments later in this chapter. As mentioned previously, the research team has also looked at the knowledge of adults and teenagers as they relate to the audience segments they fall into. Among the sobering statistics, the survey from November 2013 revealed the fact that only 5% of Americans feels that humans are going to successfully reduce global warming, while 40% or more, depending on when the survey was conducted, think that it is unclear whether or not we will. Another large segment—between one in four and one in five—are deeply pessimistic, feeling we could reduce global warming but that we won't because we aren't willing to change. Needless to say, such attitudes can't help but affect learning about these issues in school.

Other surveys, such as the 2013 study conducted by Stanford's Jon Krosnick, which conducted interviews with random Americans, found strong (75% or more) acceptance that climate change was being caused by human activities, with two-thirds calling for the United States to take action to limit greenhouse gas emissions (Nagel, 2014). Nevertheless, a dedicated and vocal minority can—and has for years—disrupt discussion and discourse, much the way one or two unruly students can derail a classroom and hamper learning. Labeling students (or adults) who doubt or dismiss climate science as deniers may be counterproductive, even though the term may indeed ring true for those who are deeply and aggressively obstinate.

In reality, denial is complicated, nuanced, and multilayered. In her article in *Time*, "We Are All Climate Change Deniers," Mary Pipher (2013) suggests that even if we do accept that climate change is occurring—and most Americans do—we tend to "minimize or normalize our enormous global problems." She writes: "Our denial is understandable. Our species is not equipped to respond to the threats posed by global warming."

In another article in *Time*—"The Battle Over Global Warming Is All in Your Head"—author Paramaguru (2013) reviews some of the psychological research that has begun to identify our mental barriers and the issues that obstruct our ability to confront the threat. Broadly defined, denial is a natural way of coping with or denying despair, a normal psychological response to cope with the angst, overwhelming feelings, or horror of a particular situation: a way to tamp down the deep despair and sense of hopelessness that may arise in contemplating catastrophe or injustice.

Pipher and sociologists like Kari Norgaard, author of *Living In Denial: Climate Change, Emotions, and Everyday Life* (2011), cite the research of Stanley Cohen, who has researched how people remain willfully ignorant about an issue out of a "need to be innocent of a troubling recognition" (p. 25). His 2001 book *States of Denial* details three primary forms of denial, all of which may come into play inside and around classrooms and other educational settings: literal (it's not happening), interpretive (it's not what you think), and implicatory (accepting the reality but denying responsibility) for what is occurring, which Pipher suggests is widespread.

While Cohen's book is about atrocities and suffering, such as genocide, and not climate change per se, his insights into how denial plays out for individuals and society also applies to human impacts on the environment in general and climate in particular. Roughly speaking, within the lens of the Six Americas segments, literal denial, though increasingly rare, is mostly often found among the Dismissive, Doubtful, and Disengaged. Interpretive denial (the planet is warming but it is because of natural cycles, not human activities) is common among these same audience segments. Implicatory denial (shirking responsibility or ignoring the implications) is arguably the most widespread, except among the most motivated of the Alarmed and Concerned.

In part because of deliberate efforts to encourage doubt and denial over the years, climate change science and potential policy solutions have become increasingly politically polarized. People who consider climate change to be overblown or a hoax derisively refer to the Alarmed group as *warmists* or *alarmists* but often take offense when they are described as *deniers*. The Heartland Institute, which has a long history of casting doubt on the health hazards of tobacco and climate science, went so far to as equate those who take the implications of climate science seriously with mass murderers and terrorists. Assailing the other side with such epithets escalates the polarization and can be counter productive if the goal is open-minded discussion and meaningful discourse. But often in the public arena, the goal has been to "win" the argument or perpetuate polarization rather than have meaningful discourse. Such polarization complicates efforts to educate people about the essential issues, leading some teachers to teach the controversy rather than the consensus science.

Efforts to offer alternative curriculum or provide cover for teaching both sides of a phony controversy often make the same three points, the three pillars of denial identified by the National Center for Science Education (http://ncse.com/climate/denial/pillars), which are the following: (1) claiming the science is bad, controversial, and/or fatally flawed; (2) suggesting that accepting the science will lead to undesirable consequences for society; and (3) insisting that therefore, for the sake of fairness and balance, both sides of the alleged controversy should be taught.

In May of 2014, a variation of this played out in the state of Wyoming when a footnote on the state's budget bill prevented implementation of the Next Generation Science Standards. The bill's author specifically called out the inclusion of climate change as the reason for preventing adoption of the standards. The reasoning? The science was bad and teaching students about climate change would destroy the state's fossil-fuel-based economy (McCaffrey, 2014). In this instance, the goal was to prevent the topic from being taught at all rather than encourage "both sides" be taught. Such political efforts to derail the teaching of climate and related energy sciences are not unique to Wyoming, and fortunately teachers and school districts are finding ways to work around political obstruction by implementing NGSS-like standards and curriculum on their own.

Denial and political polarization have had and will continue to impact whether and how climate change is taught in classrooms. The bottom line for educators is that true skepticism has a vital role in cultivating critical thinking skills in learners, but deliberate efforts to prevent teaching the topic or nit-picking designed to perpetuate endless debate and doubt need to be confronted in order to avoid furthering confusion and delay.

WHY TEACHING "BOTH SIDES" IS A PROBLEM

A first glance the idea of teaching both sides of a politically controversial topic like climate change may make sense. Many teachers, according to informal surveys conducted by the National Science Teachers Association, the Alliance for Climate Education, and the National Earth Science Teacher Association, pride themselves on teaching both sides of global warming. The reasons may vary—there may or may not be overt pressure to present the other side—but Americans' sense of fairness and balance is likely a contributing factor to the phenomenon in which educators feel that if they show a pro-climate change video, like *An Inconvenient Truth*, they are required for the sake of balance to show a video challenging climate change, like the *Great Global Warming Swindle* or *Unstoppable Solar Cycles*. In some cases well-meaning teachers will have students debate whether climate change is happening or not.

Presenting a false balance is unfair to learners because it distracts from teaching current science and can backfire, generating more confusion rather than clarity, however well intended the effort. As we observed in Chapter 2, within the context of using argumentation as a tool for delving into scientific thinking and process, good pedagogy requires argumentation only focus on genuine, contemporary scientific controversies presented at an age- and grade-appropriate level and in a reasonable scope and context.

"Both sides" false balance may also backfire by embedding a myth or misconception more deeply into someone's consciousness. When John Cook first learned from his colleague Stephan Lewandowsky about research showing how efforts to replace faulty information with correct information can backfire by over-emphasizing the myth one is trying to get rid of, he became concerned that Skeptical Science's approach to addressing climate myths was perhaps doing more harm than good.

Cook and Lewandowsky decided to pull together the relevant research into the eight-page *Debunking Handbook* (2011) that can perhaps be summed up as this: "Replace sticky thoughts with even stickier thoughts." Some of the take-home messages from the handbook include the following:

- How people think matters more than what people think.

- Complex cognitive processes are involved when refuting misinformation; it is not simply a matter of replacing a bad concept with the correct one.

- Since not mentioning a familiar myth is not always feasible, it is vital to strongly emphasize the facts you want to communicate.

- Less can be more; a long, complex explanation, even if correct, is less appealing than a simple but incorrect myth.

When individuals have a strongly held opinion or worldview, counter-arguments to correct their view may actually reinforce it. Techniques to overcome this bias include self-affirmation—expressing why a value they cherish makes them feel good—and framing—presenting the information in a way that resonates with their worldview. Finally, when a myth is effectively debunked, it is important to fill the gap with an alternative explanation. Graphics conveying core facts can be invaluable to myth busting and replacement.

There are times when repeating a sticky myth is unavoidable, but recognizing the pitfalls in calling attention to them and avoiding using debate or argumentation about faux scientific controversies can minimize the potential for doing more harm than good.

CONTEMPLATING INEQUITIES

For adults and youth in the developed world, one of the factors contributing to the varying states of denial relates to the ethical conundrum of reconciling how we, enjoying the fruits of energy-intensive lives fueled primarily by relatively abundant and inexpensive fossil fuels, are impacting everyone on the planet, especially those who did nothing to contribute to the problem. Kevin Anderson of the Tyndall Centre in the United Kingdom, one of the leading climate research institutions in the world, estimates that 20% of the world's population—primarily the developed world—are responsible for some 80% of the world's greenhouse gas emissions, and that perhaps as little as 1% of the world's population is responsible for half the emissions (2011).

But many classrooms in the U.S. may have their own energy inequities. In classes with a mix of affluent students and students in poverty, having them compare their carbon footprints may reveal wide discrepancies in energy consumption among them, which then presents a delicate and awkward teachable moment that requires sensitivity on the part of the educator.

In their article "Making Energy Access Meaningful" in *Issues in Science and Technology*, Bazilian and Pielke (2013), describing the enormous imbalance of energy consumption in the world, write "Our distinctly uncomfortable starting place is that the poorest three-quarters of the global population still only use about ten percent of global energy—a clear indicator of deep and persistent global inequity" (p.74). A question that people in the United States must ask is that, as the nation with the largest historic contributions to carbon emissions, do we have an added responsibility to prepare ourselves and the world for global changes already well underway? If so, how do we do that?

Clearly, examining economic and energy inequities and associated responsibilities are difficult and often avoided. Such topics are in many respects more appropriate in a social studies class than a science class, which is another compelling reason why climate and energy topics should be taught across the curriculum through team teaching or with teachers from various disciplines coordinating their lessons and learning goals if possible.

THE SIX AMERICAS IN THE LEARNING ENVIRONMENT

The spectrum of the Six Americas segments discussed previously offers a continuum of the relativity of acceptance, doubt, and denial. Many or all six groups will, in some form, show up in most learning environments, whether in a formal classroom, in a public outreach event, or on websites. Here are a few suggestions on how to identify and address denial and despair in these different segments.

Alarmed. Young learners in particular may feel overwhelmed by what they have learned or heard about climate change. Susie Strife, in research toward her PhD (2009), found that anger about the destruction of nature and fear of the future of the environment are common emotions felt by the ten- to twelve-year-olds she interviewed. Some had learned about environmental issues in school, but television and the Internet were the primary media that shaped their views and concerns about global warming and related topics. Video games and films of post-apocalyptic futures are common and contribute to the pessimistic and in some cases cynical attitude of some young people. Grim scenarios of an energy-constrained, substantially warmed world can indeed be alarming.

While in general people who fall into this segment are relatively more knowledgeable and willing to take action than other groups, there still may be wide variance in literacy in this segment. Some who are alarmed may be less informed on the science and more motivated by what they consider the moral or ethical imperatives of the issue than others. That said, most climate scientists who are current on the latest findings fall into this category.

Many scientists and energy experts have been trying to serve as modern-day Paul Reveres, warning of the consequences of human impacts on climate and the environment, but they have not always been effective or successful in their attempts to sound the alarm. We are now at the point where alarm and worry need to be transformed into a can-do confidence, moving beyond inaction, finger pointing, and blame. Transforming alarm into proactive action requires calmly, methodically assessing the situation and coming up with short-term and long-term strategies to apply appropriate responses: just the type of skills the Next Generation Science Standards are designed to foster.

Concerned and Cautious. These two segments, which between them account for about half of the general population, are made up of people who generally have heard of climate change, likely take it somewhat seriously, but have not made it a priority in their lives as many of the Alarmed have, feeling it isn't an immediate or pressing threat to them personally. A learning community, such as a classroom or a group of collaborating faculty, is likely to include many Concerned and Cautious. Finding ways to make climate- and energy-related issues engaging and relevant to these willing but uninvolved individuals without going overboard and pushing them into despair is the ticket.

Disengaged. While making up a small percentage of the Six Americas studies—5% in an updated, April 2013 study—in many middle and high school classrooms the percentage may be considerably higher because of apathy and/or peer pressure. The disengagement may be because of outside social factors that an educator may not be aware of or, if aware, unable to address. The disengagement, if manifested as apathy, may be in part because of the real existential quandry that climate change and related challenges pose. To reach learners who are disengaged, it is important to understand why in order to determine whether or how to move forward toward engagement.

Doubtful. Everyone has moments of doubt, but this segment, which among adults in the United States tends to be politically and/or religiously conservative, may resist attributing climate change to human activities on the grounds that humans aren't capable of altering the planet in such a way. In a science classroom, it may be possible to avoid confronting a learner's cultural background by simply saying "we're here to learn about what scientific evidence says about the planet," and if religion comes into the equation, point out that there isn't generally a conflict between religion and climate change. The National Center for Science Education's Clergy Climate Project is collecting signatures and statements from a wide range of religious leaders stating their support for the findings of climate science research and addressing the moral and ethical issues of climate change. In the case of individuals whose cultural upbringing conflicts with current climate science, encouraging an attitude of open-mindedness and inquiry may also open eyes and opportunities.

Dismissive. Determining whether someone is genuinely skeptical, in the best sense of the word, and open to learning from an individual who is locked into his or her opinion is important when dealing with those who appear to dismiss climate science. Rather than immediately jumping to the conclusion that someone is a full-fledged climate-change dissenter, it is worthwhile to reserve judgment, especially in an educational setting, to determine whether or not the person is open to learning or already has their mind made up. In some instances, the concern may be more about the danger of scaring children with alarming projections—a legitimate concern, especially with younger children, which is why age and developmentally appropriate pedagogy should always be applied.

Although a relatively small segment of society, the Dismissives have left their mark on climate education by cultivating doubt, sometimes under the guise of promoting critical thinking, using cherry-picked data points or pseudoscience. While a small minority of white males (McCright & Dunlap, 2011) may make up the core of this group, they can exert oversized influence on educators and learners by encouraging a climate of confusion and controversy.

ADDRESSING DOUBT AND DENIAL IN THE EDUCATIONAL ENVIRONMENT

In conversation, determining whether someone is earnestly trying to understand the science or is actually locked into their opinion is often very straightforward. What is the tone of the question or remark? What's your first thought on where they might fit in the Six Americas spectrum? If they are genuinely open-minded, then a thoughtful dialogue may be possible, and both parties may learn something, if not about the science involved, at least about the others' insights and perspectives.

In a science classroom of course the situation is different. While classroom management requires the educator to be firmly in the driver's seat to avoid disruptive students from hijacking the class, being sensitive to the cultural and ideological backgrounds of students is obviously vital. What about students who have family members that are convinced that climate change is a United Nations' plot to take away Americans' freedom? As many seasoned educators will attest, encouraging open-mindedness and an attitude of "let's investigate for ourselves" without directly debating conspiracy theories or criticizing students' cultural backgrounds will pay dividends. Assuming climate-related topics are in the standards or curriculum, it can also be appropriate to mention that "By the way, you may be quizzed on this down the line, and besides, it's information that will come in handy for future jobs and decisions you may have to make."

Occasionally, other teachers may undermine efforts to teach solid climate science. This may take the form of a teacher, perhaps not even a science teacher, who disparages climate scientists or Al Gore, encouraging students to be skeptical about climate change. In such instances, the best policy is to bring up the concern with someone in the school administration or, if appropriate, a union representative. If climate change is included in the official school district curriculum, teachers who have signed a contract to teach the curriculum may be in hot water if they don't teach the curriculum as laid out.

ON CONSENSUS AND UNCERTAINTY

For many years, there has been a significant disconnect between the public's view that climate scientists don't agree as to whether climate change is happening or not and the reality that there is strong agreement in the climate science community that climate is indeed changing because of human activities. Part of the confusion likely lies in the fact that there are two definitions of consensus. One is total, unanimous agreement, and the other that there is strong but not necessarily 100% agreement.

John Cook's Consensus Project (2013), similar to several other studies, has found that 97% of the papers published on climate change that take a stand on human-caused global warming agree that it is happening and humans are the cause. Therefore, those who define consensus as 100% agreement or unanimity may deny or dismiss that climate change is happening and that it is caused by human activity. Those who define consensus as strong agreement would believe the opposite, since there clearly is very robust agreement among the vast majority of peer-reviewed science papers. All major research universities, national academies, and by virtue of their ratifying the UN Framework Convention on Climate Change, 195 nations in the world agree that human activities are the driving force of climate and related global change.

Similarly, the term *uncertainty* conveys doubt and confusion among members of the public. Scientists have typically been taught to lead with their uncertainties, as a way of providing context for the evidence that follows. Often, in scientific parlance, uncertainty relates specifically to measurements and possible range of error or inaccuracy. This is similar to polling predictions where a margin of error is cited. Thus, the IPCC Fifth Assessment versus the Fourth Assessment raises the likelihood that human activities are responsible for changes in climate from 90% to 95%, meaning that scientists have gone from a 10% chance that humans aren't responsible to a 5% chance, in effect doubling the confidence. The only level higher is a 99% probability, which translates in the conservative voice of science as virtually certain. Doubting that climate change is happening or is as bad as some project can be a way to distance oneself from the ramifications and ultimately the responsibility of taking action.

PHILOSOPHIC CONUNDRUMS AND PEDAGOGICAL PRACTICES

Philosopher Stephan Gardiner, author of *A Perfect Moral Storm: The Ethical Tragedy of Climate Change*, identifies in his paper "Ethics and Climate Change: An Introduction" (2010) key areas for discussion for climate policy, which also relate to handling denial and despair, the treatment of scientific uncertainty, responsibility for past emissions, the setting of mitigation targets, and the places of adaptation and geoengineering.

In a K–12 science classroom in particular, where the focus should be on mastering the science, assigning responsibility for past emissions or discussing specific mitigation targets may be beyond the scope of the curriculum and more appropriate for other courses. But clarifying scientific uncertainty and examining adaptation and geoengineering are relevant if taught in a grade- and class-appropriate way.

The NSF-funded POLAR project at Barnard College, Columbia University—where climate change has been taught to undergraduate students for over twenty years—has developed a variety of interactive games and role-playing educational programs, including Arctic SMARTIC and Future Coast, that allow learners to think through the complex scientific and social dynamics of polar regions altered by changing climate. Dr. Stephanie Pfirman, who is the overall project lead, found that including adaptation and scenario planning for climate change up front rather than tacked on at the end of the semester resulted in more engaged students who felt inspired and empowered by the focus on things that can be done to minimize climate impacts other than reducing carbon dioxide emissions. For many years there was a concern that opening the door to adaptation planning would distract from reducing emissions, but at least in some educational settings, the proactive, anticipatory planning for impacts can be an important way to overcome despair by offering tangible things to do beyond saving energy.

Indeed, since many climate impacts take the form of natural disasters, including extreme storm events, floods, heat waves, and drought, thinking through ways of preparing for such events, doing the math on risks and probabilities, coming up with contingency plans, and engineering responses is a no-risk way of building community capacity. Such approaches also tie in strongly with the Next Generation Science Standards. The National Climate Assessment, available through http://globalchange.gov, was designed to be accessible on mobile devices, thereby taking advantage of the revolution in mobile learning, and it offers a wealth of information relating to climate adaptation throughout the United States.

Geoengineering is also a topic that many have avoided discussing or teaching as a realistic option, but given that we are currently engaged in a massive if unintended geoengineering experiment on the Earth's climate and environmental systems, the range of strategies that are being proposed to offset and counter the effect of carbon emissions on the climate system do have a place in the overall equation. Examining these strategies may also provide opportunities for learners to delve into cutting-edge science, technology, engineering, and mathematics to weigh their practical—and ethical—pros and cons. The majority of the proposals are fraught with ethical, political, and practical issues, particularly the potential for unintended consequences. The two primary strategies are

- Solar radiation management schemes, including cloud modification and altering albedo and land-cover on the Earth's surface or building a sun-shade in space to reduce incoming solar energy; and

- Greenhouse gas remediation other than reducing emissions, including carbon capture and sequestration, air-capture through chemical processes, ocean fertilization with iron or urea, and biochar—a form of charcoal that, when buried, sequesters carbon in the ground, improves soil fertility, and acts as a carbon filter for groundwater recharge.

Because geoengineering covers such a range of options, some far-fetched, some inherently low-tech and practical, it cannot be immediately written off. That said, in an educational setting, such as a science classroom, the topic should be introduced in a grade- or course-appropriate way, since learners should understand the basic climate system and how humans are impacting the system before they tackle an analysis of possible solutions.

Klaus Lackner, Director of the Lenfest Center for Sustainable Energy at Columbia University, works on carbon capture and sequestration strategies but was inspired by his daughter's eighth-grade science fair project that extracted carbon dioxide from the air using a fish pump and sodium hydroxide. See the PBS video of Dr. Lackner's project, which is included in the CLEAN collection: http://cleanet.org/resources/43035.html. Other projects, such as Global Thermostat—http://globalthermostat.com—which uses thermal processes to capture carbon dioxide, and New Sky Energy—http://www.newskyenergy.com—which extracts it from wastewater, offer examples of emerging entrepreneurial opportunities that take advantage of cutting-edge science.

DEALING WITH DESPAIR

In his comments about "where to go from here," Jorg Friedrichs, author of *The Future Is Not What It Used to Be* (2013), suggests that we may well be heading for a hard landing when even well-informed people are unable or unwilling to confront what he describes as "the transitory nature of industrial society." His solution is for the moral individual to live "in the truth" because:

> Life is tragic and sometimes there are no solutions . . . Insofar as climate change and energy scarcity are part of the human predicament, even the most

accurate diagnosis is unlikely to suggest an easy cure. And yet, my mission as a scholar is to get to the bottom of things regardless of whether or not there is a solution. This does not mean that, as a citizen and consumer, I am better than anyone else. My task as a scholar is not to save the planet or pose as an ecological do-gooder. It is plain old-fashioned intellectual honesty. (p. 170)

He acknowledges that many readers will find his conclusions depressing, especially those who believe the problems can easily remedied simply through politics or local activism. Hoping he is ultimately proven wrong, he suggests two tools—resilience thinking and preventing loss of self-identity—that may prove helpful. The first will require rethinking how we harness energy and in a sense reinventing the goodness of our humanity through our values. He is convinced, however, that sustaining the status quo, just making the current system more resilient won't work: More needs to be done.

Friedrich suggests that emergency measures need to be taken to prevent loss of self-identity, which is vital for individuals to have in order to successfully engage in society and help transform it. Loss of identity occurs on every level of society, and young people, who are in the process of forming their self-identify, are vulnerable to becoming alienated from others and from their environment unless they are given the knowledge to both know themselves and know the challenges facing the future. Arguably, many people numb themselves with video games, substances, and consuming because the pain of facing the world as it is is too much to bear. Overcoming apathy, doubt, and denial in order to make the topic come alive as interesting and relevant is vitally important.

Not all learners will necessarily be troubled by negative emotions or despair. Individuals can be remarkably resilient even in the most challenging circumstances, accepting the "facts of life" and then asking, "What do I do about it?" This opens the door to taking action, however small or tentative that first step might be.

In a sense, data about climate and energy are neutral—just numbers. The context and implications are not: Humanity faces massive challenges on every level, and pessimism is a natural response for many of us. Confronting the despair, cutting through the denial, recognizing that there are options, and focusing on interdisciplinary education and life-long learning—inherently optimistic enterprises if ever there were ones—gives us practical tools and visionary strategies for the future. Taking action is an important part of that future, and potentially part of the fun, as we'll see in Chapter 8.

ADDITIONAL RESOURCES

Biochar, International. (n.d.). International Biochar Initiative website. Retrieved from http://www.bio char-international.org

> *This organization offers an interdisciplinary and integrating theme that allows learners to explore many facets of the carbon cycle. The International Biochar Initiative has begun to collect examples of student projects.*

CHAPTER 8

Knowledge, Know-How, and Informed Action

How can we prepare our young people, the nation, and by extension the world to tackle the climate and energy challenges of the 21st century through improved literacy and learning environments? One way would be through a large-scale campaign set in motion through a public-private partnership, ideally working in parallel with local and regional partnerships that are focused on two ambitious and audacious goals:

1. Providing the 76 million students in the nation—about 56 million K–12 students and another 20 million in higher education and professional programs—with the knowledge that they will need to make informed climate and energy decisions in their lives as well as to minimize risks and maximize resiliency in their communities

2. Transforming the 140,000 schools in the nation (100,000 public K–12, approximately 33,000 private, and some 7,000 higher education and professional schools, many with multiple buildings or campuses) into living laboratories of innovation and learning that are climate-smart, energy-wise, disaster-ready centers for learners and the communities they serve

These two goals are complementary, and we know there are many examples of success in both areas to point to: from effective, engaging curriculum at every grade level that can serve as a source of inspiration for teachers and students, to schools large and small that have transformed their learning environments into hotbeds of creativity for sustainable living, preparing young people today for the workforce and world of tomorrow.

The National Climate Assessment Network (NCAnet), a public-private partnership of over 100 organizations that came together to support the National Climate Assessment, offers the scaffolding on which to build such an initiative, and the assessment itself is a robust resource of actionable science that can be unpacked in educational settings to identify responses as well as related job opportunities and career pathways.

The reason for focusing on literacy and education is simple. In any given year, roughly one out of every four people in the United States is a student, and millions more are parents, grandparents, employers, or future employers. Until recently, most students never learned the basics of climate and global change or the causes, effects, and possible responses to them. This has contributed enormously to the climate of confusion on these topics in the United States that has been deliberately perpetuated by vested interests. Ideally, learning should occur in a safe, inspiring environment, which is precisely why educational facilities must be upgraded to reflect the state-of-the-art, 21st-century building design.

As Stiglitz and Greenwald emphasize in their book *Creating a Learning Society* (2014, p. 62), education is pivotal to cultivating the "learning mindset" necessary for a flourishing society, hence it must be a priority and not an afterthought or add-on to addressing the climate and energy challenges we face.

CULTIVATING SEEDS OF SUCCESS

Some seeds of success to achieve this vision have already been planted. The informal science education sphere of museums and science centers have expertise on how to make these vital topics interactive and compelling to learners of all ages. High quality climate, energy, and risk-reduction educational materials already exist, including those recommended throughout this book, but widespread distribution of them is needed.

Schools and school districts have conducted detailed energy audits, gaining awareness of the climate control challenges within their buildings and proposing ways to deal with those challenges. The Next Generation Science Standards, which include energy as a crosscutting theme, address minimizing risks and negative impacts through a solutions-focused lens linked to other areas of the curriculum. The standards have started to gain acceptance and application in the education community.

A few years ago, climate change typically wasn't taught at all, but preliminary results from a survey conducted for the Understanding Global Change project at the University of California at Berkeley indicate this is beginning to change (McCaffrey, Berbeco, White, & Stuhlsatz, 2013). Most science teachers in the survey, who were self-selected and may not be representative of all teachers, reported they do at least touch on climate change and feel it is a priority to teach. New studies are underway to learn more about whether, where, and how climate and global change topics, including energy, ecosystems, and sustainability, are taught.

Some K–12 and higher education schools have developed and begun implementing sustainability and climate action plans. As was cited in Chapter 6, a variety of green school programs currently exist, from the U.S. Department of Education's Green Ribbon Schools award program to the international Green School Alliance. In higher education, the American College and University Presidents' Climate Commitment has served as a catalyst, with students, faculty, staff, and administrators collaborating to reduce human impacts and transform their learning institutions.

Efforts have been set in motion at every level of society to upgrade schools and make them healthier and more efficient. The U.S. Green Building Council, for example, tracks and encourages energy efficiency in schools and has found very robust support for the idea of green schools nationwide (Shelter, 2013). In California, funding through

Proposition 39, which closed a corporate loophole, is putting hundreds of millions of dollars toward reducing energy waste in public schools, creating many teachable moments. A 12.1 megawatt solar project involving 28,000 panels at schools throughout the district has allowed the Mt. Diablo Unified School District in California to generate over 90% of the energy needed for the district, saving more than $3 million a year (Barnidge, 2014), opening up numerous "teachable moments" in the schools where the solar arrays are deployed.

When it comes to preparing for and reducing risks from natural or human-caused disasters, both the U.S. Department of Education and the Department of Homeland Security offer guidance for schools on how to prepare for and reduce potential risks. But many schools lack comprehensive disaster plans and are themselves vulnerable, even though, when emergencies arise, the go-to sites for people in harm's way are often schools.

PUBLIC-PRIVATE PARTNERSHIPS: MAXIMIZING POTENTIAL ENERGY

What has been lacking is a coordinated and sustained campaign to link these components together. Current efforts have been piecemeal, short-term, and often poorly funded. But with a long-term focus and coordination, within a decade schools can not only become community R&D hubs for learning and innovation but also potential sources of distributed energy generation and/or grid batteries for energy storage. Located in the majority of communities across the nation, our schools serve as voting locations and refuges during emergencies, but they could also serve as nodes for powering—and thereby empowering—their communities.

Because the scale and scope of the endeavor and our shared responsibility to do everything we can to prepare and protect this and future generations from natural and human-caused hazards and risks, a large-scale strategic partnership is required, bringing together federal, state, and local agencies, nonprofit and for profit organizations, and philanthropic interests. The National Climate Assessment Network (NCAnet) or a similar public-private partnership could serve as the backbone supporting the coordination of such an initiative. Key to the success of such an endeavor will be effective facilitation that works toward consensus while encouraging action at every level of society commensurate with the urgency of the energy and climate challenges the world faces.

To maximize the collective success of a public-private partnership mobilized to achieve these goals, a long-term strategic process for development and implementation is necessary. Process participants need to do the following:

Years 1–2

- Begin and broaden the conversation.

- Develop a baseline of where and how climate, energy, and risk preparedness/ reduction are being taught.

- Identify effective practices and exemplary models that can be transferred and adopted.

- Conduct an inventory of the nation's school building stock, focusing on relative energy efficiency and resilience to risks.

- Build community and community capacity to collaborate and act.

Years 3–5

- Continue and refine the conversation.

- Prioritize education, infrastructure, and readiness needs.

- Fund worthy projects and conduct evaluation to refine and improve effectiveness.

- Provide substantial professional development for educators around these interrelated topics, stressing the community-scale citizen science opportunities that can be led by youth and guided by experts.

Years 5–10

- Enrich and empower the conversation.

- Methodically and strategically transform the nation's schools into interlinked hubs of learning and innovation and distributed power generation.

- Share the best of the climate-smart, energy-wise, disaster-ready living laboratories throughout the nation and around the world.

- Respond quickly and effectively to the inevitable disasters.

There are clearly innovative programs to make school facilities and surrounding infrastructure, like transportation, more energy efficient, using renewables like solar and geothermal where appropriate, but many schools lack the financial wherewithal to make these investments. Even when such efficiencies are added, the link to curriculum is often missing, losing valuable teachable moments for students and the community in the process. Whether through a large-scale initiative or more localized efforts, to fully transform the curriculum and the schools themselves to be truly climate smart, energy wise, and disaster ready will require tackling the problems from every angle: top-down, bottom-up, and especially inside out. Every teacher and student, parent and administrator has an important role to play in advocating for significant climate and energy content in the curriculum and for schools to become living laboratories of energy efficiency and innovation.

A national public-private partnership would provide incentives, track progress, and remove barriers at the national, state, and school district level. With the shared vision of 100% literacy about the essentials of climate and energy within a generation and the aim that all schools be safe, efficient learning and energy hubs for the communities they serve, we can in relatively short order ready ourselves, our children, and future generations for the knowns and unknowns of the future.

Education is inherently local, so there is often understandable pushback to intrusive testing or requirements that impede local control. The tension between national, state, and local agendas will certainly need to be kept front and center in such a public-private partnership. Ultimately, decisions do need to be made at the local building and district level about how best to meet the needs of the community and its learners. This is where students, parents, teachers, and other local stakeholders will likely find the greatest impact, starting conversations and working toward solutions.

There is clearly a role for district and state leaders to determine how best to deploy curriculum, prepare teachers, and prioritize infrastructure upgrades and repairs. There is a role for mayors and other local leaders to help contribute to the process, and many of them are already heavily involved with climate action plans to minimize impacts and risks. There is a role for businesses and foundations that have a stake in building community resilience and student preparedness. And there is also a role for the federal government. For example, there are currently nearly 20 million K–12 students who receive free or reduced lunch in the United States. The vast majority of funding to help insure students receive basic nutrition while at school comes from the federal government through the states. But the federal government also has a responsibility to serve and protect all of its citizens from threats, whether foreign or domestic, naturally occurring or human induced, which is one reason the U.S. Department of Defense takes climate disruption seriously as a "threat multiplier." All the 13 federal agencies involved with the U.S. Global Change Research Program either directly protect the nation or conduct important research that helps inform decision making.

What if a national scale public-private partnership doesn't emerge to mobilize the transformation of climate-smart, energy-wise schools for the 21st century? There is still much that can be done on a smaller scale to advocate for improved and expanded teaching and upgraded, safer schools. Beginning locally is always a good place to begin, and any national-scale initiative will have to be customized to meet local needs.

Whether or not a national-scale public-private partnership gains traction, there is much that can be done at every level of society to work toward a more climate-smart, energy-wise world. Most importantly, we need to

- **Start the conversations.** Teachers can talk with their fellow faculty and administrators, and parents and students can speak with teachers and administrators to voice their views.

- **Forge partnerships.** None of us can transform our knowledge and know-how systems (i.e., our schools) singlehandedly. Once the conversation is underway, we must look for partners who we share something in common with. In some cases we may have little else in common with them other than an abiding interest in the future of our children, which may be enough to get started.

- **Pick our battles.** Most of us have limited time and energy, and our interests may not be toward lobbying for funding for improved infrastructure or for students' right to learn about climate and energy topics. But we can find our niche that aligns with our interests and skills.

INFORMED CLIMATE AND ENERGY ACTION

The "Guiding Principle for Climate Literacy," which was reviewed and endorsed by 13 federal agencies that are part of the U.S. Global Change Research Program, is remarkably and refreshingly explicit: Humans can take actions to reduce climate change and its impacts. This "Guiding Principle for Informed Climate Decisions," included in Appendix II, focuses on actionable science, describing how climate science improves policy and decision making, how reducing human vulnerability to and impacts on climate requires multidisciplinary, integrated understanding, and how climate change affects national security. It also examines how greenhouse gas reduction and carbon dioxide sequestration mitigate climate change and cites strategies to reduce greenhouse gas emission (energy conservation, renewable energies, change in energy use) as well as strategies for humans adapting to climate change. Finally, it emphasizes that actions taken by different levels of society can mitigate climate change and increase preparedness for current and future generations.

Does this amount to advocacy? In the sense that it advocates for using science and rationality to inform our decisions about climate and related challenges, yes, it does. But the "Guiding Principle for Informed Climate Decisions" is not proscriptive in terms of promoting a particular policy or behavior change. This is an important distinction.

The word "advocacy" has sometimes been framed as something to be avoided in science education in particular. But clearly we should be advocates for learners and do everything in our power to prepare and protect our nation's learners for global change, transforming our schools to be climate smart and safe, energy wise and efficient. Yes, we must provide the care and feeding of a true learning society capable of tackling the daunting challenges of the coming years and decades. We need to be advocates for informed decision making and actions based on the evidence and its implications.

And, yes, we should advocate teaching these topics throughout the grade levels, ideally infusing climate and energy topics throughout the curriculum—not only in the physical, life, and Earth sciences, but also in mathematics, language arts, social studies, civics, and arts. This increases flexibility and allows educators to better shape curriculum and priorities on a local level and to customize strategies to meet the specific needs of their learners. In a civics or social studies class, for example, it may be appropriate to have students learn about and respond to a particular issue, but in a science class, learning the science needs to be the focus. Such a cross-disciplinary, collaborative approach broadens and deepens the learning.

Naturally, some well-meaning educators go overboard, foisting their opinions on students. Having students write letters to Congress, for example, or encouraging them to become vegan because the teacher is vegan can be problematic, leading to charges of indoctrination. Although academic freedom in higher education gives teachers more free rein to profess their views to students, in the K–12 realm, where delivering the required curriculum is challenging enough, this can become an issue. As Minda Berbeco wrote in her post *Action in Climate Education: A Step Too Far?*:

> To be clear, students absolutely must learn the science, understand the causes
> and know the potential solutions. How they act on that knowledge, though,

is a personal decision that, for better or worse, cannot be dictated by an educator. The teacher's role is to educate, not to proselytize. (2013)

A Few Final Words

As with the revolution in learning inspired by the Enlightenment values of science and rationality that led to the American Revolution and many vital societal transformations since, the current literacy revolution that is now happening in schools and communities around the nation to provide young people with the knowledge and know-how to address the climate, energy, socioeconomic, and ecological challenges presents an all-hands-on-deck situation similar to the mobilization and transformation of society during and after the Second World War. Each of us ideally has our own unique intelligence, creativity, insights, passion, and willingness to bring to this learning revolution that will prepare the nation and the people of the planet for changes that are already well underway.

When people ask me, "What can I do about climate change?" (often followed by, "Besides changing light bulbs or buying a Prius"), the answer I give is usually along these lines: Do everything you possibly can to become informed about the science and surrounding societal context. Understand the causes, effects, and risks of global change and know the range of responses. The National Climate Assessment (http://nca2014 .globalchange.gov), which contains a wide array of actionable science, is a good starting place. Then do your utmost to prepare and protect this and future generations from climate and other global change and energy challenges by supporting climate, energy, and global change literacy and informed decision making. Get involved in helping transform schools and thereby communities. Challenge yourself and find that sweet spot between being under- and overwhelmed.

We must all find our niche and then widen it, realize our potential and energize it, be our own light and then link it up in series with others', cultivate our own proverbial garden and then join with others to build our community's gardens. We must transform the current climate of confusion into one of caring and clarity.

Appendix I

Voices for Climate Education

U.S. GOVERNMENT

U.S. Global Change Research Program (USGCRP)

People want to know how climate change and variability affect their lives and livelihoods and whether humans can and should do something about global warming. There is a growing imperative for the actionable climate and environmental information that is needed to inform resource management, planning, and other decisions taking place across the nation.

http://www.globalchange.gov/what-we-do/communication-and-education

NATIONAL RESEARCH COUNCIL/NATIONAL ACADEMIES

Climate Change Education: Goals, Audiences, and Strategies (2011)

The reality of global climate change lends increasing urgency to the need for effective education on Earth system sciences, as well as on the human and behavioral dimensions of climate change, from broad societal action to smart energy choices at the household level.

http://www.nap.edu/openbook.php?record_id=13224&page=1

Informing an Effective Response to Climate Change (2010)

The climate-related decisions that society will confront over the coming decades will require an informed and engaged public and an education system that provides students with the knowledge they need to make informed choices about responses to climate change. Today's students will become tomorrow's decision makers as business leaders, farmers, government officials, and citizens. Our report finds that much more could be done to improve climate literacy, increase public understanding of climate science and choices, and inform decision makers about climate change, including an urgent need for research on effective methods of climate change education and communication.

http://www.nap.edu/openbook.php?record_id=12784&page=1

EDUCATIONAL ORGANIZATIONS

National Association of Geoscience Teachers

The National Association of Geoscience Teachers (NAGT) recognizes: (1) that Earth's climate is changing, (2) that present warming trends are largely the result of human activities, and (3) that teaching climate change science is a fundamental and integral part of Earth science education.

http://nagt.org/nagt/policy/ps-climate.html

PROFESSIONAL SCIENCE ORGANIZATIONS

American Geophysical Union

Actions that could diminish the threats posed by climate change to society and ecosystems include substantial emissions cuts to reduce the magnitude of climate change, as well as preparing for changes that are now unavoidable. The community of scientists has responsibilities to improve overall understanding of climate change and its impacts. Improvements will come from pursuing the research needed to understand climate change, working with stakeholders to identify relevant information, and conveying understanding clearly and accurately, both to decision makers and to the general public.

http://sciencepolicy.agu.org/files/2013/07/AGU-Climate-Change-Position-Statement_August-2013.pdf

Geological Society of America

The Geological Society recommends that its members take the following actions:

> Actively participate in professional education and discussion activities to be technically informed about the latest advances in climate science. . . . Engage in public education activities in the community, including the local level.

http://www.geosociety.org/positions/position10.htm

American Chemical Society

[The ACS recommends the development of] a national strategy to support climate change education and communication that both involves students, technical professionals, public servants, and the general public, as well as being integrated with state and local initiatives. A national climate education act could serve as a catalyzing agent to reinvigorate science, technology, engineering, and mathematics (STEM) education across the nation.

http://www.acs.org/content/acs/en/policy/publicpolicies/promote/globalclimatechange.html

SOURCE: http://ncse.com/climate/taking-action/voices-climate-change-education

Appendix II

Excerpts From *Climate Literacy: The Essential Principles of Climate Science*

A Climate-Oriented Approach for Learners of All Ages

WHAT IS CLIMATE SCIENCE LITERACY?

Climate Science Literacy is an understanding of your influence on climate and climate's influence on you and society. A climate-literate person

- understands the essential principles of Earth's climate system,

- knows how to assess scientifically credible information about climate,

- communicates about climate and climate change in a meaningful way, and

- is able to make informed and responsible decisions with regard to actions that may affect climate.

CLIMATE SCIENCE LITERACY IS AN ONGOING PROCESS

No single person is expected to understand every detail about all of the fundamental climate science literacy concepts. Full comprehension of these interconnected concepts will require a systems-thinking approach, meaning the ability to understand complex interconnections among all of the components of the climate system. Moreover, as climate science progresses and as efforts to educate the people about climate's influence on them and their influence on the climate system mature, public understanding will continue to grow.

Climate is an ideal interdisciplinary theme for lifelong learning about the scientific process and the ways in which humans affect and are affected by the Earth's systems. This rich topic can be approached at many levels, from comparing the daily weather with long-term records to exploring abstract representations of climate in computer models to examining how climate change impacts human and ecosystem health. Learners of all ages can use data from their own experiments, data collected by satellites and other observation systems, or records from a range of physical, chemical, biological, geographical, social, economic, and historical sources to explore the impacts of climate and potential adaptation and mitigation strategies.

How Do We Know What Is Scientifically Correct?

THE PEER REVIEW PROCESS

Science is an ongoing process of making observations and using evidence to test hypotheses. As new ideas are developed and new data are obtained, oftentimes enabled by new technologies, our understanding evolves. The scientific community uses a highly formalized version of peer review to validate research results and our understanding of their significance. Researchers describe their experiments, results, and interpretations in scientific manuscripts and submit them to a scientific journal that specializes in their field of science. Scientists who are experts in that field serve as referees for the journal: they read the manuscript carefully to judge the reliability of the research design and check that the interpretations are supported by the data. Based on the reviews, journal editors may accept or reject manuscripts or ask the authors to make revisions if the study has insufficient data or unsound interpretations. Through this process, only those concepts that have been described through well-documented research and subjected to the scrutiny of other experts in the field become published papers in science journals and accepted as current science knowledge. Although peer review does not guarantee that any particular published result is valid, it does provide a high assurance that the work has been carefully vetted for accuracy by informed experts prior to publication. The overwhelming majority of peer-reviewed papers about global climate change acknowledge that human activities are substantially contributing factors.

Guiding Principle for Informed Climate Decisions

Humans can take actions to reduce climate change and its impacts.

A. Climate information can be used to reduce vulnerabilities or enhance the resilience of communities and ecosystems affected by climate change. Continuing to improve scientific understanding of the climate system and the quality of reports to policy and decision-makers is crucial.

B. Reducing human vulnerability to the impacts of climate change depends not only upon our ability to understand climate science, but also upon our ability to integrate that knowledge into human society. Decisions that involve Earth's climate must be made with an understanding of the complex inter-connections among the physical and biological components of the Earth system as well as the consequences of such decisions on social, economic, and cultural systems.

C. The impacts of climate change may affect the security of nations. Reduced availability of water, food, and land can lead to competition and conflict among humans, potentially resulting in large groups of climate refugees.

D. Humans may be able to mitigate climate change or lessen its severity by reducing greenhouse gas concentrations through processes that move carbon out of the atmosphere or reduce greenhouse gas emissions.

E. A combination of strategies is needed to reduce greenhouse gas emissions. The most immediate strategy is conservation of oil, gas, and coal, which we rely on as fuels for most of our transportation, heating, cooling, agriculture, and electricity. Short-term strategies involve switching from carbon-intensive to renewable energy sources, which also requires building new infrastructure for alternative energy sources. Long-term strategies involve innovative research and a fundamental change in the way humans use energy.

F. Humans can adapt to climate change by reducing their vulnerability to its impacts. Actions such as moving to higher ground to avoid rising sea levels, planting new crops that will thrive under new climate conditions, or using new building technologies represent adaptation strategies. Adaptation often requires financial investment in new or enhanced research, technology, and infrastructure.

G. Actions taken by individuals, communities, states, and countries all influence climate. Practices and policies followed in homes, schools, businesses, and governments can affect climate. Climate-related decisions made by one generation can provide opportunities as well as limit the range of possibilities open to the next generation. Steps toward reducing the impact of climate change may influence the present generation by providing other benefits such as improved public health infrastructure and sustainable built environments.

CLIMATE CHANGES

Throughout its history, Earth's climate has varied, reflecting the complex interactions and dependencies of the solar, oceanic, terrestrial, atmospheric, and living components that make up planet Earth's systems. For at least the last million years, our world has experienced cycles of warming and cooling that take approximately 100,000 years to complete. Over the course of each cycle, global average temperatures have fallen and then risen again by about 9°F (5°C), each time taking Earth into an ice age and then warming it again. This cycle is believed to be associated with regular changes in Earth's orbit that alter the intensity of solar energy the planet receives. Earth's climate has also been influenced on very long timescales by changes in ocean circulation that result from plate tectonic movements. Earth's climate has changed abruptly at times, sometimes as a result of slower natural processes such as shifts in ocean circulation, sometimes due to sudden events such as massive volcanic eruptions. Species and ecosystems have either adapted to these past climate variations or perished.

While global climate has been relatively stable over the last 10,000 years—the span of human civilization—regional variations in climate patterns have influenced human history in profound ways, playing an integral role in whether societies thrived or failed. We now know that the opposite is also true: human activities—burning fossil fuels and deforesting large areas of land, for instance—have had a profound influence on Earth's climate. In its 2007 Fourth Assessment, the Intergovernmental Panel on Climate Change (IPCC) stated that it had "very high confidence that the global average net effect of human activities since 1750 has been one of warming." The IPCC attributes humanity's global warming influence primarily to the increase in three key heat-trapping gases in the atmosphere: carbon dioxide, methane, and nitrous oxide. The U.S. Global Change Research Program published findings in agreement with the IPCC report, stating that "studies to detect climate change and attribute its causes using patterns of observed temperature change in space and time show clear evidence of human influences on the climate system (due to changes in greenhouse gases, aerosols, and stratospheric ozone)."[1]

To protect fragile ecosystems and to build sustainable communities that are resilient to climate change—including extreme weather and climate events—a climate-literate citizenry is essential. This *climate science literacy* guide identifies the essential principles and fundamental concepts that individuals and communities should understand about Earth's climate system. Such understanding improves our ability to make decisions about activities that increase vulnerability to the impacts of climate change and to take precautionary steps in our lives and livelihoods that would reduce those vulnerabilities.

KEY DEFINITIONS

Adaptation   Initiatives and measures to reduce the vulnerability of natural and human systems against actual or expected climate change effects.[2]

Carbon Cycle   Circulation of carbon atoms through the Earth systems as a result of photosynthetic conversion of carbon dioxide into complex organic compounds by plants, which are consumed by other organisms, and return of the carbon to the atmosphere as carbon dioxide as a result of respiration, decay of organisms, and combustion of fossil fuels.

Climate Change   A significant and persistent change in the mean state of the climate or its variability. Climate change occurs in response to changes in some aspect of Earth's environment: these include regular changes in Earth's orbit about the sun, re-arrangement of continents through plate tectonic motions, or anthropogenic modification of the atmosphere.

Climate Forecast   A prediction about average or extreme climate conditions for a region in the long-term future (seasons to decades).

Climate System   The matter, energy, and processes involved in interactions among Earth's atmosphere, hydrosphere, cryosphere, lithosphere, biosphere, and Earth-Sun interactions.

Climate Variability   Natural changes in climate that fall within the normal range of extremes for a particular region, as measured by temperature, precipitation, and frequency of events. Drivers of climate variability include the El Niño Southern Oscillation and other phenomena.

Climate   The long-term average of conditions in the atmosphere, ocean, and ice sheets and sea ice described by statistics, such as means and extremes.

Feedback   The process through which a system is controlled, changed, or modulated in response to its own output. Positive feedback results in amplification of the system output; negative feedback reduces the output of a system.

Fossil Fuels   Energy sources such as petroleum, coal, or natural gas, which are derived from living matter that existed during a previous geologic time period.

Global Warming   The observed increase in average temperature near the Earth's surface and in the lowest layer of the atmosphere. In common usage, global warming often refers to the warming that has occurred as a result of increased emissions of greenhouse gases from human activities. Global warming is a type of climate change; it can also lead to other changes in climate conditions, such as changes in precipitation patterns.

Likely, Very Likely, Extremely Likely, Virtually Certain   These terms are used by the Intergovernmental Panel on Climate Change (IPCC) to indicate how probable it is that a predicted outcome will occur in the climate system, according to expert judgment. A result that is deemed "likely" to occur has a greater than 66% probability of occurring. A "very likely" result has a greater than 90% probability. "Extremely likely" means greater than 95% probability, and "virtually certain" means greater than 99% probability.

Mitigation   Human interventions to reduce the sources of greenhouse gases or enhance the sinks that remove them from the atmosphere.

Vulnerability   The degree to which physical, biological, and socio-economic systems are susceptible to and unable to cope with adverse impacts of climate change.[3]

Weather Forecast   A prediction about the specific atmospheric conditions expected for a location in the short-term future (hours to days).

Weather   The specific conditions of the atmosphere at a particular place and time, measured in terms of variables that include temperature, precipitation, cloudiness, humidity, air pressure, and wind.

NOTES

1. *Temperature Trends in the Lower Atmosphere: Steps for Understanding and Reconciling Differences.* Thomas R. Karl, Susan J. Hassol, Christopher D. Miller, and William L. Murray, editors, 2006. A Report by the Climate Change Science Program and the Subcommittee on Global Change Research, Washington, DC.

2. Based on IPCC, 2007: *Mitigation of Climate Change. Contribution of Working Group III*

3. Based on IPCC, 2007: *Impacts, Adaptation and Vulnerability. Contribution of Working Group II*

SOURCES: http://globalchange.gov/resources/educators/climate-literacy.html and http://cpo.noaa.gov/OutreachandEducation/ClimateLiteracy.aspx

CLIMATE LITERACY: THE ESSENTIAL PRINCIPLES OF CLIMATE SCIENCE

1. THE SUN IS THE PRIMARY SOURCE OF ENERGY FOR EARTH'S CLIMATE SYSTEM.

A. Sunlight reaching the Earth can heat the land, ocean, and atmosphere. Some of that sunlight is reflected back to space by the surface, clouds, or ice. Much of the sunlight that reaches Earth is absorbed and warms the planet.

B. When Earth emits the same amount of energy as it absorbs, its energy budget is in balance, and its average temperature remains stable.

C. The tilt of Earth's axis relative to its orbit around the Sun results in predictable changes in the duration of daylight and the amount of sunlight received at any latitude throughout a year. These changes cause the annual cycle of seasons and associated temperature changes.

D. Gradual changes in Earth's rotation and orbit around the Sun change the intensity of sunlight received in our planet's polar and equatorial regions. For at least the last 1 million years, these changes occurred in 100,000-year cycles that produced ice ages and the shorter warm periods between them.

E. A significant increase or decrease in the Sun's energy output would cause Earth to warm or cool. Satellite measurements taken over the past 30 years show that the Sun's energy output has changed only slightly and in both directions. These changes in the Sun's energy are thought to be too small to be the cause of the recent warming observed on Earth.

2. CLIMATE IS REGULATED BY COMPLEX INTERACTIONS AMONG COMPONENTS OF THE EARTH SYSTEM.

A. Earth's climate is influenced by interactions involving the Sun, ocean, atmosphere, clouds, ice, land, and life. Climate varies by region as a result of local differences in these interactions.

B. Covering 70% of Earth's surface, the ocean exerts a major control on climate by dominating Earth's energy and water cycles. It has the capacity to absorb large amounts of solar energy. Heat and water vapor are redistributed globally through density-driven ocean currents and atmospheric circulation. Changes in ocean circulation caused by tectonic movements or large influxes of fresh water from melting polar ice can lead to significant and even abrupt changes in climate, both locally and on global scales.

C. The amount of solar energy absorbed or radiated by Earth is modulated by the atmosphere and depends on its composition. Greenhouse gases—such as water vapor, carbon dioxide, and methane—occur naturally in small amounts and absorb and release heat energy more efficiently than abundant atmospheric gases like nitrogen and oxygen. Small increases in carbon dioxide concentration have a large effect on the climate system.

D. The abundance of greenhouse gases in the atmosphere is controlled by biogeochemical cycles that continually move these components between their ocean, land, life, and atmosphere reservoirs. The abundance of carbon in the atmosphere is reduced through seafloor accumulation of marine sediments and accumulation of plant biomass and is increased through deforestation and the burning of fossil fuels as well as through other processes.

E. Airborne particulates, called aerosols, have a complex effect on Earth's energy balance: they can cause both cooling, by reflecting incoming sunlight back out to space, and warming, by absorbing and releasing heat energy in the atmosphere. Small solid and liquid particles can be lofted into the atmosphere through a variety of natural and man-made processes, including volcanic eruptions, sea spray, forest fires, and emissions generated through human activities.

F. The interconnectedness of Earth's systems means that a significant change in any one component of the climate system can influence the equilibrium of the entire Earth system. Positive feedback loops can amplify these effects and trigger abrupt changes in the climate system. These complex interactions may result in climate change that is more rapid and on a larger scale than projected by current climate models.

3. LIFE ON EARTH DEPENDS ON, IS SHAPED BY, AND AFFECTS CLIMATE.

A. Individual organisms survive within specific ranges of temperature, precipitation, humidity, and sunlight. Organisms exposed to climate conditions outside their normal range must adapt or migrate, or they will perish.

B. The presence of small amounts of heat-trapping greenhouse gases in the atmosphere warms Earth's surface, resulting in a planet that sustains liquid water and life.

C. Changes in climate conditions can affect the health and function of ecosystems and the survival of entire species. The distribution patterns of fossils show evidence of gradual as well as abrupt extinctions related to climate change in the past.

D. A range of natural records shows that the last 10,000 years have been an unusually stable period in Earth's climate history. Modern human societies developed during this time. The agricultural, economic, and transportation systems we rely upon are vulnerable if the climate changes significantly.

E. Life—including microbes, plants, and animals and humans—is a major driver of the global carbon cycle and can influence global climate by modifying the chemical makeup of the atmosphere. The geologic record shows that life has significantly altered the atmosphere during Earth's history.

4. CLIMATE VARIES OVER SPACE AND TIME THROUGH BOTH NATURAL AND MAN-MADE PROCESSES.

A. Climate is determined by the long-term pattern of temperature and precipitation averages and extremes at a location. Climate descriptions can refer to areas that are local, regional, or global in extent. Climate can be described for different time intervals, such as decades, years, seasons, months, or specific dates of the year.

B. Climate is not the same thing as weather. Weather is the minute-by-minute variable condition of the atmosphere on a local scale. Climate is a conceptual description of an area's average weather conditions and the extent to which those conditions vary over long time intervals.

C. Climate change is a significant and persistent change in an area's average climate conditions or their extremes. Seasonal variations and multi-year cycles (for example, the El Niño Southern Oscillation) that produce warm, cool, wet, or dry periods across different regions are a natural part of climate variability. They do not represent climate change.

D. Scientific observations indicate that global climate has changed in the past, is changing now, and will change in the future. The magnitude and direction of this change is not the same at all locations on Earth.

E. Based on evidence from tree rings, other natural records, and scientific observations made around the world, Earth's average temperature is now warmer than it has been for at least the past 1,300 years. Average temperatures have increased markedly in the past 50 years, especially in the North Polar Region.

F. Natural processes driving Earth's long-term climate variability do not explain the rapid climate change observed in recent decades. The only explanation that is consistent with all available evidence is that human impacts are playing an increasing role in climate change. Future changes in climate may be rapid compared to historical changes.

G. Natural processes that remove carbon dioxide from the atmosphere operate slowly when compared to the processes that are now adding it to the atmosphere. Thus, carbon dioxide introduced into the atmosphere today may remain there for a century or more. Other greenhouse gases, including some created by humans, may remain in the atmosphere for thousands of years.

5. OUR UNDERSTANDING OF THE CLIMATE SYSTEM IS IMPROVED THROUGH OBSERVATIONS, THEORETICAL STUDIES, AND MODELING.

A. The components and processes of Earth's climate system are subject to the same physical laws as the rest of the Universe. Therefore, the behavior of the climate system can be understood and predicted through careful, systematic study.

B. Environmental observations are the foundation for understanding the climate system. From the bottom of the ocean to the surface of the Sun, instruments on weather stations, buoys, satellites, and other platforms collect climate data. To learn about past climates, scientists use natural records, such as tree rings, ice cores, and sedimentary layers. Historical observations, such as native knowledge and personal journals, also document past climate change.

C. Observations, experiments, and theory are used to construct and refine computer models that represent the climate system and make predictions about its future behavior. Results from these models lead to better understanding of the linkages between the atmosphere-ocean system and climate conditions and inspire more observations and experiments. Over time, this iterative process will result in more reliable projections of future climate conditions.

D. Our understanding of climate differs in important ways from our understanding of weather. Climate scientists' ability to predict climate patterns months, years, or decades into the future is constrained by different limitations than those faced by meteorologists in forecasting weather days to weeks into the future.[1]

E. Scientists have conducted extensive research on the fundamental characteristics of the climate system and their understanding will continue to improve. Current climate change projections are reliable enough to help humans evaluate potential decisions and actions in response to climate change.

6. HUMAN ACTIVITIES ARE IMPACTING THE CLIMATE SYSTEM.

A. The overwhelming consensus of scientific studies on climate indicates that most of the observed increase in global average temperatures since the latter part of the 20th century is very likely due to human activities, primarily from increases in greenhouse gas concentrations resulting from the burning of fossil fuels.[2]

B. Emissions from the widespread burning of fossil fuels since the start of the Industrial Revolution have increased the concentration of greenhouse gases in the atmosphere. Because these gases can remain in the atmosphere for hundreds of years before being removed by natural processes, their warming influence is projected to persist into the next century.

C. Human activities have affected the land, oceans, and atmosphere, and these changes have altered global climate patterns. Burning fossil fuels, releasing chemicals into the atmosphere, reducing the amount of forest cover, and rapid expansion of farming, development, and industrial activities are releasing carbon dioxide into the atmosphere and changing the balance of the climate system.

D. Growing evidence shows that changes in many physical and biological systems are linked to human-caused global warming.[3] Some changes resulting from human activities have decreased the capacity of the environment to support various species and have substantially reduced ecosystem biodiversity and ecological resilience.

E. Scientists and economists predict that there will be both positive and negative impacts from global climate change. If warming exceeds 2 to 3°C (3.6 to 5.4°F) over the next century, the consequences of the negative impacts are likely to be much greater than the consequences of the positive impacts.

7. CLIMATE CHANGE WILL HAVE CONSEQUENCES FOR THE EARTH SYSTEM AND HUMAN LIVES.

A. Melting of ice sheets and glaciers, combined with the thermal expansion of seawater as the oceans warm, is causing sea level to rise. Seawater is beginning to move onto low-lying land and to contaminate coastal fresh water sources and beginning to submerge coastal facilities and barrier islands. Sea-level rise increases the risk of damage to homes and buildings from storm surges such as those that accompany hurricanes.

B. Climate plays an important role in the global distribution of freshwater resources. Changing precipitation patterns and temperature conditions will alter the distribution and availability of freshwater resources, reducing reliable access to water for many people and their crops. Winter snowpack and mountain glaciers that provide water for human use are declining as a result of global warming.

C. Incidents of extreme weather are projected to increase as a result of climate change. Many locations will see a substantial increase in the number of heat waves they experience per year and a likely decrease in episodes of severe cold. Precipitation events are expected to become less frequent but more intense in many areas, and droughts will be more frequent and severe in areas where average precipitation is projected to decrease.[2]

D. The chemistry of ocean water is changed by absorption of carbon dioxide from the atmosphere. Increasing carbon dioxide levels in the atmosphere is causing ocean water to become more acidic, threatening the survival of shell-building marine species and the entire food web of which they are a part.

E. Ecosystems on land and in the ocean have been and will continue to be disturbed by climate change. Animals, plants, bacteria, and viruses will migrate to new areas with favorable climate conditions. Infectious diseases and certain species will be able to invade areas that they did not previously inhabit.

F. Human health and mortality rates will be affected to different degrees in specific regions of the world as a result of climate change. Although cold-related deaths are predicted to decrease, other risks are predicted to rise. The incidence and geographical range of climate-sensitive infectious diseases—such as malaria, dengue fever, and tick-borne diseases—will increase. Drought-reduced crop yields, degraded air and water quality, and increased hazards in coastal and low-lying areas will contribute to unhealthy conditions, particularly for the most vulnerable populations.[3]

NOTES

1. Based on Climate Change: An Information Statement of the American Meteorological Society, 2007

2. Based on IPCC, 2007: *The Physical Science Basis: Contribution of Working Group I*

3. Based on IPCC, 2007: *Impacts, Adaptation and Vulnerability. Contribution of Working Group II*

SOURCE: U.S. Department of Energy

Appendix III

Excerpts From *Energy Literacy: Essential Principles and Fundamental Concepts for Energy Education*

WHAT IS ENERGY LITERACY?

Energy literacy is an understanding of the nature and role of energy in the universe and in our lives. Energy literacy is also the ability to apply this understanding to answer questions and solve problems.

An energy-literate person:

- can trace energy flows and think in terms of energy systems

- knows how much energy he or she uses, for what, and where the energy comes from

- can assess the credibility of information about energy

- can communicate about energy and energy use in meaningful ways

- is able to make informed energy and energy use decisions based on an understanding of impacts and consequences

- continues to learn about energy throughout his or her life

WHY DOES ENERGY LITERACY MATTER?

A better understanding of energy can:

- lead to more informed decisions

- improve the security of a nation

- promote economic development

- lead to sustainable energy use

- reduce environmental risks and negative impacts

- help individuals and organizations save money

Without a basic understanding of energy, energy sources, generation, use, and conservation strategies, individuals and communities cannot make informed decisions on topics ranging from smart energy use at home and consumer choices to national and international energy policy. Current national and global issues such as the fossil fuel supply and climate change highlight the need for energy education.

A Brief History of Human Energy Use

Producers in a food chain—like plants, algae and cyanobacteria—capture energy from the Sun. Nearly all organisms rely on this energy for survival. Energy flow through most food chains begins with this captured solar energy. Some of this energy is used by organisms at each level of the food chain, much is lost as heat, and a small portion is passed down the food chain as one organism eats another.

Over time, humans have developed an understanding of energy that has allowed them to harness it for uses well beyond basic survival.

The first major advance in human understanding of energy was the mastery of fire. The use of fire to cook food and heat dwellings, using wood as the fuel, dates back at least 400,000 years.[1] The burning of wood and other forms of biomass eventually led to ovens for making pottery, and the refining of metals from ore. The first evidence of coal being burned as a fuel dates back approximately 2,400 years.[2]

After the advent of fire, human use of energy per capita remained nearly constant until the Industrial Revolution of the 19th century. This is despite the fact that, shortly after mastering fire, humans learned to use energy from the Sun, wind, water, and animals for endeavors such as transportation, heating, cooling, and agriculture.

The invention of the steam engine was at the center of the Industrial Revolution. The steam engine converted the chemical energy stored in wood or coal into motion energy. The steam engine was widely used to solve the urgent problem of pumping water out of coal mines. As improved by James Watt, Scottish inventor and mechanical engineer, it was soon used to move coal, drive the manufacturing of machinery, and power locomotives, ships and even the first automobiles.[3] It was during this time that coal replaced wood as the major fuel supply for industrialized society. Coal remained the major fuel supply until the middle of the 20th century when it was overtaken by oil.

The next major energy revolution was the ability to generate electricity and transmit it over large distances. During the first half of the 19th century, British physicist Michael Faraday demonstrated that electricity would flow in a wire exposed to a changing magnetic field, now known as Faraday's Law. Humans then understood how to generate electricity. In the 1880s, Nikola Tesla, a Serbian-born electrical engineer, designed alternating current (AC) motors and transformers that made long-distance transmission of electricity possible. Humans could now generate electricity on a large scale, at a single location, and then transmit that electricity efficiently to many different locations. Electricity generated at Niagara Falls, for example, could be used by customers all over the region.

Although hydropower, largely in the form of water wheels, has been in use by human society for centuries, hydroelectricity is a more recent phenomenon. The first hydroelectric power plants were built at the end of the 19th century and by the middle of the 20th century were a major source of electricity. As of 2010, hydropower produced more than 15% of the world's electricity.[4]

Humans have also been using energy from wind to power human endeavors for centuries, but have only recently begun harnessing wind energy to generate electricity. Wind energy propelled boats along the Nile River as early as 5000 B.C. By 200 B.C., simple windmills in China were pumping water, while vertical-axis windmills with woven reed sails were grinding grain in Persia and the Middle East. Windmills designed to generate electricity, or wind turbines, appeared in Denmark as early as 1890. Currently, wind provides almost 2% of the world's electricity.[5]

In the 20th century, Einstein's Theories of Relativity and the new science of quantum mechanics brought with them an understanding of the nature of matter and energy that gave rise to countless new technologies. Among these technologies were the nuclear power plant and the solar or photovoltaic cell. Both of these technologies emerged as practical sources of electricity in the 1950s. Nuclear energy quickly caught on as a means of generating electricity. Today, nuclear energy generates almost 15% of the world's electricity. Solar energy provides less than 1% of the world's electricity. Solar is the only primary energy source that can generate electricity without relying on Faraday's Law. Particles of light can provide the energy for the flow of electrons directly.

Humans have also managed to harness the geothermal energy of Earth to produce electricity. The first geothermal power plant was built in 1911 in Larderello, Italy. Geothermal energy is a result of the continuous radioactive decay of unstable elements beneath Earth's surface and gravitational energy associated with Earth's mass. The radioactive decay and gravitational energy produce thermal energy that makes its way to the surface of Earth, often in the form of hot water or steam.

Modern biofuels are another way humans have found to harness energy for use beyond basic survival. Biofuels are plant materials and animal waste used as fuel. For example, ethanol is a plant-based fuel used more and more commonly in vehicles, usually in conjunction with petroleum-based fuels.

Although humans have found many different sources of energy to power their endeavors, fossil fuels remain the major source by a wide margin. The three fossil fuel sources are coal, oil, and natural gas. Oil has been the major fuel supply for industrialized society since the middle of the 20th century and provides more of the energy used by humans than any other source. Coal is second on this list, followed closely by natural gas. Together they accounted for more than 80% of the world's energy use in 2010.

Industrialization and the rise in access to energy resources have taken place at very disparate rates in different countries around the world. For example, as of 2011, there were 1.3 billion people on Earth with no access to electricity.[6]

As with any human endeavor, the use of energy resources and the production of electricity have had and will continue to have impacts and consequences, both good and bad. Awareness of the energy used to grow, process, package, and transport food, or the energy used to treat water supplies and wastewater, is important if society is to minimize waste and maximize efficiency. These are just a few examples of the many energy issues people can become informed about.

Human society has and will continue to develop rules and regulations to help minimize negative consequences. As new information comes to light and new technologies are developed, energy policies are reevaluated, requiring individuals and communities to make decisions. This guide outlines the understandings necessary for these decisions to be informed.

FIGURE III.1   U.S. Primary Energy Consumption Estimates by Source, 1775-2010

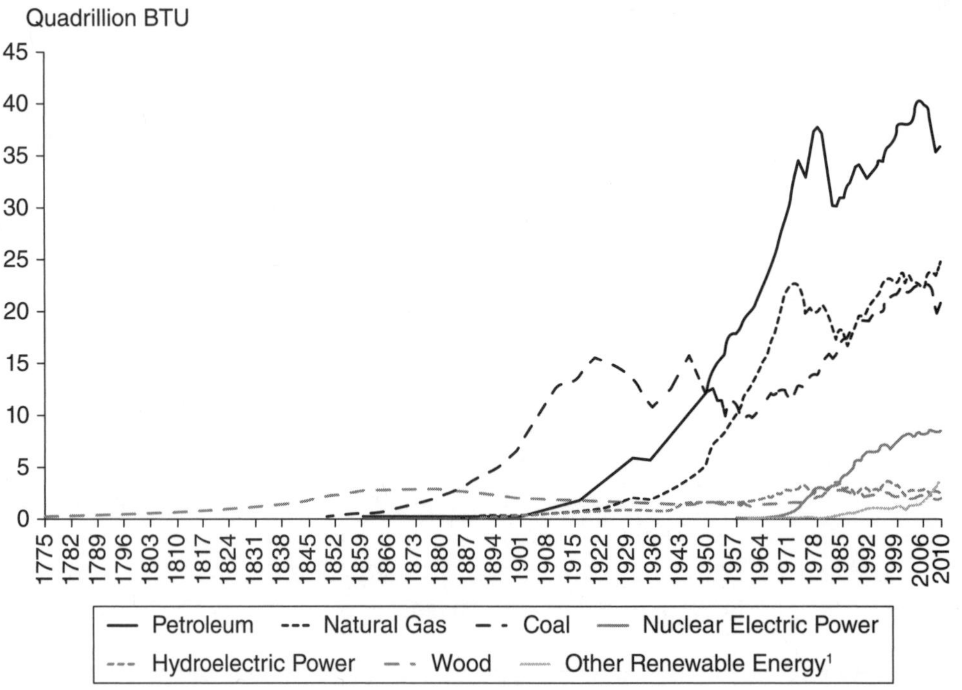

SOURCE: U.S. Energy Information Administration Annual Energy Review, Tables 1.3, 10.1, and E1. http://www.eia.gov/totalenergy/data/annual/perspectives.cfm

NOTE: 1. Geothermal, solar/PV, wind, waste, and biofuels.

Notes

1. Bowman DM, Balch JK, Artaxo P et al. Fire in the Earth System. Science. 2009, 324 (5926), pp 481–4.
2. Metalworking and Tools, in: Oleson, John Peter (ed.): The Oxford Handbook of Engineering and Technology in the Classical World, Oxford University Press, 2009, pp. 418–38 (432).
3. Benchmarks for Science Literacy, American Association for the Advancement of Science, 1993, benchmark 10J/M2.
4. Source of data is the U.S. Energy Information Administration (http://www.eia.gov) unless otherwise noted
5. World Wind Energy Report, World Wind Energy Association, February 2009.
6. International Energy Agency, World Energy Outlook, 2011.

Available at http://www1.eere.energy.gov/education/energy_literacy.html

ENERGY LITERACY

THE ESSENTIAL PRINCIPLES AND FUNDAMENTAL CONCEPTS

A note on the use of the Essential Principles and Fundamental concepts:

The Essential Principles, 1 through 7, are meant to be broad categories representing big ideas. Each Essential Principle is supported by six to eight Fundamental Concepts: 1.1, 1.2, and so on. The Fundamental Concepts are intended to be unpacked and applied as appropriate for the learning audience and setting. For example, teaching about the various sources of energy (Fundamental Concept 4.1) in a 3rd grade classroom, in a 12th grade classroom, to visitors of a museum, or as part of a community education program will look very different in each case. Further, the concepts are not intended to be addressed in isolation; a given lesson on energy will most often connect to many of the concepts.

1. Energy is a physical quantity that follows precise natural laws.

2. Physical processes on Earth are the result of energy flow through the Earth system.

3. Biological processes depend on energy flow through the Earth system.

4. Various sources of energy can be used to power human activities, and often this energy must be transferred from source to destination.

5. Energy decisions are influenced by economic, political, environmental, and social factors.

6. The amount of energy used by human society depends on many factors.

7. The quality of life of individuals and societies is affected by energy choices.

1. Energy is a physical quantity that follows precise natural laws.

1.1 **Energy is a quantity that is transferred from system to system.** Energy is the ability of a system to do work. A system has done work if it has exerted a force on another system over some distance. When this happens, energy is transferred from one system to another. At least some of the energy is also transformed from one type into another during this process. One can keep track of how much energy transfers into or out of a system.

1.2 **The energy of a system or object that results in its temperature is called thermal energy.** When there is a net transfer of energy from one system to another, due to a difference in temperature, the energy transferred is called heat. Heat transfer happens in three ways: convection, conduction, and radiation. Like all energy transfer, heat transfer involves forces exerted over a distance at some level as systems interact.

1.3 **Energy is neither created nor destroyed.** The change in the total amount of energy in a system is always equal to the difference between the amount of energy transferred in and the amount transferred out. The total amount of energy in the universe is finite and constant.

1.4 Energy available to do useful work decreases as it is transferred from system to system. During all transfers of energy between two systems, some energy is lost to the surroundings. In a practical sense, this lost energy has been used up, even though it is still around somewhere. A more efficient system will lose less energy, up to a theoretical limit.

1.5 Energy comes in different forms and can be divided into categories. Forms of energy include light energy, elastic energy, chemical energy, and more. There are two categories that all energy falls into: kinetic and potential. Kinetic describes types of energy associated with motion. Potential describes energy possessed by an object or system due to its position relative to another object or system and forces between the two. Some forms of energy are part kinetic and part potential energy.

1.6 Chemical and nuclear reactions involve transfer and transformation of energy. The energy associated with nuclear reactions is much larger than that associated with chemical reactions for a given amount of mass. Nuclear reactions take place at the centers of stars, in nuclear bombs, and in both fission- and fusion-based nuclear reactors. Chemical reactions are pervasive in living and non-living Earth systems.

1.7 Many different units are used to quantify energy. As with other physical quantities, many different units are associated with energy. For example, joules, calories, ergs, kilowatt-hours, and BTUs are all units of energy. Given a quantity of energy in one set of units, one can always convert it to another (e.g., 1 calorie = 4.186 joules).

1.8 Power is a measure of energy transfer rate. It is useful to talk about the rate at which energy is transferred from one system to another (energy per time). This rate is called power. One joule of energy transferred in one second is called a Watt (i.e., 1 joule/second = 1 Watt).

2. Physical processes on Earth are the result of energy flow through the Earth system.

2.1 Earth is constantly changing as energy flows through the system. Geologic, fossil, and ice records provide evidence of significant changes throughout Earth's history. These changes are always associated with changes in the flow of energy through the Earth system. Both living and non-living processes have contributed to this change.

2.2 Sunlight, gravitational potential, decay of radioactive isotopes, and rotation of the Earth are the major sources of energy driving physical processes on Earth. Sunlight is a source external to Earth, while radioactive isotopes and gravitational potential, with the exception of tidal energy, are internal. Radioactive isotopes and gravity work together to produce geothermal energy beneath Earth's surface. Earth's rotation influences global flow of air and water.

2.3 Earth's weather and climate are mostly driven by energy from the Sun. For example, unequal warming of Earth's surface and atmosphere by the Sun drives convection within the atmosphere, producing winds, and influencing ocean currents.

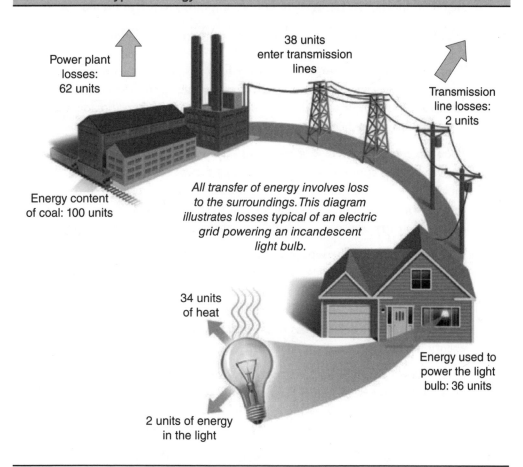

Power plant losses: 62 units

38 units enter transmission lines

Transmission line losses: 2 units

Energy content of coal: 100 units

All transfer of energy involves loss to the surroundings. This diagram illustrates losses typical of an electric grid powering an incandescent light bulb.

34 units of heat

Energy used to power the light bulb: 36 units

2 units of energy in the light

SOURCE: Reprinted with permission from *What You Need to Know About Energy*, 2008, National Academy of Sciences and National Academies Press.

2.4 Water plays a major role in the storage and transfer of energy in the Earth system. The major role water plays is a result of water's prevalence, high heat capacity, and the fact that phase changes of water occur regularly on Earth. The Sun provides the energy that drives the water cycle on Earth.

2.5 Movement of matter between reservoirs is driven by Earth's internal and external sources of energy. These movements are often accompanied by a change in the physical and chemical properties of the matter. Carbon, for example, occurs in carbonate rocks such as limestone, in the atmosphere as carbon dioxide gas, in water as dissolved carbon dioxide, and in all organisms as complex molecules that control the chemistry of life. Energy drives the flow of carbon between these different reservoirs.

2.6 Greenhouse gases affect energy flow through the Earth system. Greenhouse gases in the atmosphere, such as carbon dioxide and water vapor, are transparent to much of the incoming sunlight but not to the infrared light from the warmed surface of Earth. These gases play a major role in determining average global surface temperatures. When Earth emits the same amount of energy as it absorbs, its average temperature remains stable.

2.7 **The effects of changes in Earth's energy system are often not immediately apparent.** Responses to changes in Earth's energy system, input versus output, are often only noticeable over the course of months, years, or even decades.

3. Biological processes depend on energy flow through the Earth system.

3.1 **The Sun is the major source of energy for organisms and the ecosystems of which they are a part.** Producers such as plants, algae, and cyanobacteria use the energy from sunlight to make organic matter from carbon dioxide and water. This establishes the beginning of energy flow through almost all food webs.

3.2 **Food is a biofuel used by organisms to acquire energy for internal living processes.** Food is composed of molecules that serve as fuel and building material for all organisms as energy stored in the molecules is released and used. The breakdown of food molecules enables cells to store energy in new molecules that are used to carry out the many functions of the cell and thus the organism.

3.3 **Energy available to do useful work decreases as it is transferred from organism to organism.** The chemical elements that make up the molecules of living things are passed through food chains and are combined and recombined in different ways. At each level in a food chain, some energy is stored in newly made chemical structures, but most is dissipated into the environment. Continual input of energy, mostly from sunlight, keeps the process going.

3.4 **Energy flows through food webs in one direction, from producers to consumers and decomposers.** An organism that eats lower on a food chain is more energy efficient than one eating higher on a food chain. Eating producers is the lowest, and thus most energy efficient, level at which an animal can eat.

3.5 **Ecosystems are affected by changes in the availability of energy and matter.** The amount and kind of energy and matter available constrains the distribution and abundance of organisms in an ecosystem and the ability of the ecosystem to recycle materials.

3.6 **Humans are part of Earth's ecosystems and influence energy flow through these systems.** Humans are modifying the energy balance of Earth's ecosystems at an increasing rate. The changes happen, for example, as a result of changes in agricultural and food processing technology, consumer habits, and human population size.

4. Various sources of energy can be used to power human activities, and often this energy must be transferred from source to destination.

4.1 **Humans transfer and transform energy from the environment into forms useful for human endeavors.** The primary sources of energy in the environment include fuels like coal, oil, natural gas, uranium, and biomass. All primary source fuels except biomass are non-renewable. Primary sources also include renewable sources such as sunlight, wind, moving water, and geothermal energy.

4.2 **Human use of energy is subject to limits and constraints.** Industry, transportation, urban development, agriculture, and most other human activities are closely tied to the amount and kind of energy available. The availability of energy resources is constrained by the distribution of natural resources, availability of affordable technologies, socioeconomic policies, and socioeconomic status.

FIGURE III.3 • Producers to Consumers: Energy Flow Through Trophic Levels

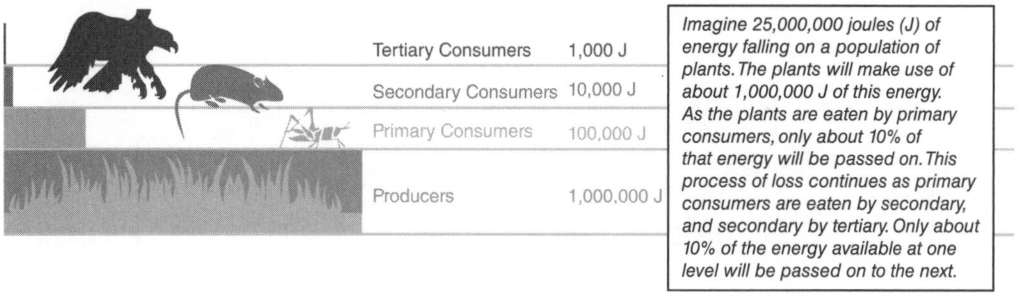

Tertiary Consumers	1,000 J
Secondary Consumers	10,000 J
Primary Consumers	100,000 J
Producers	1,000,000 J

Imagine 25,000,000 joules (J) of energy falling on a population of plants. The plants will make use of about 1,000,000 J of this energy. As the plants are eaten by primary consumers, only about 10% of that energy will be passed on. This process of loss continues as primary consumers are eaten by secondary, and secondary by tertiary. Only about 10% of the energy available at one level will be passed on to the next.

SOURCE: U.S. Department of Energy.

4.3 Fossil and biofuels are organic matter that contain energy captured from sunlight. The energy in fossil fuels such as oil, natural gas, and coal comes from energy that producers such as plants, algae, and cyanobacteria captured from sunlight long ago. The energy in biofuels such as food, wood, and ethanol comes from energy that producers captured from sunlight very recently. Energy stored in these fuels is released during chemical reactions, such as combustion and respiration, which also release carbon dioxide into the atmosphere.

4.4 Humans transport energy from place to place. Fuels are often not used at their source but are transported, sometimes over long distances. Fuels are transported primarily by pipelines, trucks, ships, and trains. Electrical energy can be generated from a variety of energy resources and can be transformed into almost any other form of energy. Electric circuits are used to distribute energy to distant locations. Electricity is not a primary source of energy, but an energy carrier.

4.5 Humans generate electricity in multiple ways. When a magnet moves or magnetic field changes relative to a coil of wire, electrons are induced to flow in the wire. Most human generation of electricity happens in this way. Electrons can also be induced to flow through direct interaction with light particles; this is the basis upon which a solar cell operates. Other means of generating electricity include electrochemical, piezoelectric, and thermoelectric.

4.6 Humans intentionally store energy for later use in a number of different ways. Examples include batteries, water reservoirs, compressed air, hydrogen, and thermal storage. Storage of energy involves many technological, environmental, and social challenges.

4.7 Different sources of energy and the different ways energy can be transformed, transported, and stored each have different benefits and drawbacks. A given energy system, from source to sink, will have an inherent level of energy efficiency, monetary cost, and environmental risk. Each system will also have national security, access, and equity implications.

FIGURE III.4   U.S. Primary Energy Consumption by Energy Source, 2011

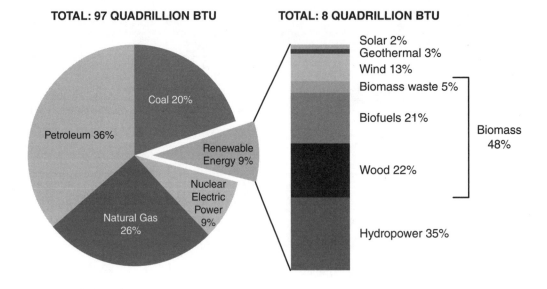

TOTAL: 97 QUADRILLION BTU

TOTAL: 8 QUADRILLION BTU

Coal 20%

Petroleum 36%

Renewable Energy 9%

Nuclear Electric Power 9%

Natural Gas 26%

Solar 2%
Geothermal 3%
Wind 13%
Biomass waste 5%
Biofuels 21%
Wood 22%
Biomass 48%
Hydropower 35%

SOURCE: U.S. Energy Information Administration, Annual Energy Review 2010.

NOTE: Sum of biomass components does not equal 53% due to independent rounding.

5. Energy decisions are influenced by economic, political, environmental, and social factors.

5.1 Decisions concerning the use of energy resources are made at many levels. Humans make individual, community, national, and international energy decisions. Each of these levels of decision making has some common and some unique aspects. Decisions made beyond the individual level often involve a formally established process of decision-making.

5.2 Energy infrastructure has inertia. The decisions that governments, corporations, and individuals made in the past have created today's energy infrastructure. The large amount of money, time, and technology invested in these systems makes changing the infrastructure difficult, but not impossible. The decisions of one generation both provide and limit the range of possibilities open to the future generations.

5.3 Energy decisions can be made using a systems-based approach. As individuals and societies make energy decisions, they can consider the costs and benefits of each decision. Some costs and benefits are more obvious than others. Identifying all costs and benefits requires a careful and informed systems-based approach to decision making.

5.4 Energy decisions are influenced by economic factors. Monetary costs of energy affect energy decision making at all levels. Energy exhibits characteristics of both a commodity and a differentiable product. Energy costs are often subject to market fluctuations, and energy choices made by individuals and societies affect these fluctuations. Cost differences also arise as a result of differences between energy sources and as a result of tax-based incentives and rebates.

5.5 Energy decisions are influenced by political factors. Political factors play a role in energy decision making at all levels. These factors include, but are not limited

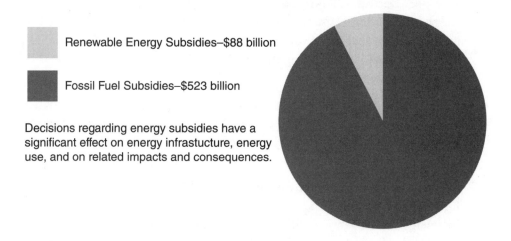

FIGURE III.5   Global Energy Subsidies, 2011

Renewable Energy Subsidies—$88 billion

Fossil Fuel Subsidies—$523 billion

Decisions regarding energy subsidies have a significant effect on energy infrastucture, energy use, and on related impacts and consequences.

SOURCE: International Energy Agency (IEA), World Energy Outlook, 2012

to, governmental structure and power balances, actions taken by politicians, and partisan-based or self-serving actions taken by individuals and groups.

5.6   Energy decisions are influenced by environmental factors. Environmental costs of energy decisions affect energy decision making at all levels. All energy decisions have environmental consequences. These consequences can be positive or negative.

5.7   Energy decisions are influenced by social factors. Questions of ethics, morality, and social norms affect energy decision making at all levels. Social factors often involve economic, political, and environmental factors.

6. The amount of energy used by human society depends on many factors.

6.1   Conservation of energy has two very different meanings. There is the physical law of conservation of energy. This law says that the total amount of energy in the universe is constant. Conserving energy is also commonly used to mean the decreased use of societal energy resources. When speaking of people conserving energy, this second meaning is always intended.

6.2   One way to manage energy resources is through conservation. Conservation includes reducing wasteful energy use, using energy for a given purpose more efficiently, making strategic choices as to sources of energy, and reducing energy use altogether.

6.3   Human demand for energy is increasing. Population growth, industrialization, and socioeconomic development result in increased demand for energy. Societies have choices with regard to how they respond to this increase. Each of these choices has consequences.

6.4   Earth has limited energy resources. Increasing human energy consumption places stress on the natural processes that renew some energy resources and it depletes those that cannot be renewed.

6.5 Social and technological innovation affects the amount of energy used by human society. The amount of energy society uses per capita or in total can be decreased. Decreases can happen as a result of technological or social innovation and change. Decreased use of energy does not necessarily equate to decreased quality of life. In many cases it will be associated with increased quality of life in the form of increased economic and national security, reduced environmental risks, and monetary savings.

6.6 Behavior and design affect the amount of energy used by human society. There are actions individuals and society can take to conserve energy. These actions might come in the form of changes in behavior or in changes to the design of technology and infrastructure. Some of these actions have more impact than others.

6.7 Products and services carry with them embedded energy. The energy needed for the entire lifecycle of a product or service is called the embedded or embodied energy. An accounting of the embedded energy in a product or service, along with knowledge of the source(s) of the energy, is essential when calculating the amount of energy used and in assessing impacts and consequences.

6.8 Amount of energy used can be calculated and monitored. An individual, organization, or government can monitor, measure, and control energy use in many ways. Understanding utility costs, knowing where consumer goods and food come from, and understanding energy efficiency as it relates to home, work, and transportation are essential to this process.

FIGURE III.6   Where Does My Money Go?

Annual energy bill for a typical U.S. single family home is approximately $2,200.

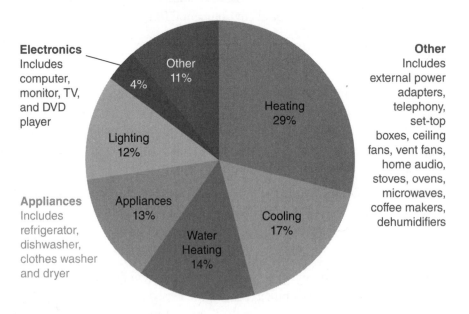

Electronics
Includes computer, monitor, TV, and DVD player

Appliances
Includes refrigerator, dishwasher, clothes washer and dryer

Other
Includes external power adapters, telephony, set-top boxes, ceiling fans, vent fans, home audio, stoves, ovens, microwaves, coffee makers, dehumidifiers

Other 11%
Heating 29%
4%
Lighting 12%
Appliances 13%
Water Heating 14%
Cooling 17%

SOURCE: Typical House memo, Lawrence Berkeley National Laboratory, 2009 and Typical house_2009_Reference.xls spreadsheet

NOTES: Average price of electricity is 11.3 cents per kilo-watt hour. Average price of natural gas is $13.29 per million BTU.

CLIMATE SMART & ENERGY WISE

7. The quality of life of individuals and societies is affected by energy choices.

7.1 Economic security is impacted by energy choices. Individuals and society continually make energy choices that have economic consequences. These consequences come in the form of monetary cost in general and in the form of price fluctuation and instability specifically.

7.2 National security is impacted by energy choices. The security of a nation is dependent, in part, on the sources of that nation's energy supplies. For example, a nation that has diverse sources of energy that come mostly from within its borders is more secure than a nation largely dependent on foreign energy supplies.

7.3 Environmental quality is impacted by energy choices. Energy choices made by humans have environmental consequences. The quality of life of humans and other organisms on Earth can be significantly affected by these consequences.

7.4 Increasing demand for and limited supplies of fossil fuels affects quality of life. Fossil fuels provide the vast majority of the world's energy. Fossil fuel supplies are limited. If society has not transitioned to sources of energy that are renewable before depleting Earth's fossil fuel supplies, it will find itself in a situation where energy demand far exceeds energy supply. This situation will have many social and economic consequences.

7.5 Access to energy resources affects quality of life. Access to energy resources, or lack thereof, affects human health, access to education, socioeconomic status, gender equality, global partnerships, and the environment.

7.6 Some populations are more vulnerable to impacts of energy choices than others. Energy decisions have economic, social, and environmental consequences. Poor, marginalized, or underdeveloped populations can most benefit from positive consequences and are the most susceptible to negative consequences.

KEY DEFINITIONS

Definitions given here are for the purposes of this document and are not necessarily complete or exhaustive. Words or phrases included here are those for which there may be some confusion as to the meaning intended.

Biofuel A fuel produced from biomass or biomass used directly as a fuel. Compare Biomass.

Biomass Organic non-fossil material of biological origin. Compare Biofuel.

Chemical Reaction A process that involves changes in the structure and energy content of atoms, molecules, or ions but not their nuclei. Compare Nuclear Reaction.

Commodity A good for which there is demand, but which is supplied without qualitative differentiation across a market. The market treats it as equivalent or nearly so no matter who produces it. Compare Differentiable Product.

Conservation of Energy See Fundamental Concept 6.1.

Degrade (as in energy)   The transformation of energy into a form in which it is less available for doing work.

Differentiable Product   A product whose price is not universal. A product whose price is based on factors such as brand and perceived quality. Compare Commodity.

Efficient   The use of a relatively small amount of energy for a given task, purpose, or service; achieving a specific output with less energy input.

Embedded or Embodied Energy   See Fundamental Concept 6.7.

Energy Carrier   A source of energy that has been subject to human-induced energy transfers or transformations. Examples include hydrogen fuel and electricity. Compare Primary Energy Source.

Energy   See Fundamental Concept 1.1.

Fossil Fuel   Fuel formed from biomass by a process taking millions of years or longer.

Fuel   A material substance that possesses internal energy that can be transferred to the surroundings for specific uses. Included are petroleum, coal, and natural gas (the fossil fuels), and other materials, such as uranium, hydrogen, and biofuels.

Geothermal Energy   See Fundamental Concept 2.2.

Heat   See Fundamental Concept 1.2.

Kinetic Energy   See Fundamental Concept 1.5.

Nuclear Reaction   A reaction, as in fission, fusion, or radioactive decay, that alters the energy, composition, or structure of an atomic nucleus. Compare Chemical Reaction.

Political   Of, relating to, or dealing with the structure or affairs of government, politics, or the state. Or, relating to, involving characteristic of politics or politicians. Or, based on or motivated by partisan or self-serving objectives.

Potential Energy   See Fundamental Concept 1.5.

Power   See Fundamental Concept 1.8.

Primary Energy Source or Primary Source   A source of energy found in nature that has not been subject to any human-induced energy transfers or transformations. Examples include fossil fuels, solar, wind and hydropower. Compare Energy Carrier.

Renewable Energy   Energy obtained from sources that are virtually inexhaustible (defined in terms of comparison to the lifetime of the Sun) and replenish naturally over small time scales relative to the human life span.

Reservoir   A place where a supply or store of something is kept or located.

Sustainable   Able to be maintained at a steady level without exhausting natural resources or causing severe ecological damage, as in a behavior or practice.

System   A set of connected things or parts forming a complex whole. In particular, a set of things working together as parts of a mechanism or an interconnecting network. The place one system ends and another begins is not an absolute, but instead must be defined based on purpose and situation.

Systems-Based Approach   An approach that emphasizes the interdependence and interactive nature of elements within and external to events, processes, and phenomena. An approach that seeks to identify and understand all cause-and-effect connections related to a given event, process, or phenomenon.

Thermal Energy   See Fundamental Concept 1.2.

Work   See Fundamental Concept 1.1.

SOURCE: U.S. Department of Energy

References

American Association for the Advancement of Science. (1993). *Benchmarks for science literacy.* New York, NY: University of Oxford Press.

American Geosciences Institute. (2013). *Earth and space sciences education in U.S. secondary schools: Key indicators and trends.* (Earth and Space Sciences Report No.1:0). Retrieved from http://geocntr.org/wpcontent/uploads/2013/08/ESS_sec_status_report_10_17_13.pdf

Anderson, A. (2010). *Combating climate change through quality education.* (Policy Brief 2010–2013). Washington, DC: Brookings Institution. Retrieved from http://www.brookings.edu/~/media/research/files/papers/2010/9/climate%20education%20anderson/09_climate_education.pdf

Anderson, K. (2011, July). *Climate change: Going beyond dangerous . . . brutal numbers & tenuous hope or cognitive dissonance?* Slideshow presented at Tyndall Centre, University of Manchester. Retrieved from http://www.slideshare.net/DFID/professor-kevin-anderson-climate-change-going-beyond-dangerous

Arrhenius, S. (1896, April). On the influence of carbonic acid in the air upon the temperature of the ground. *Philosophical Magazine and Journal of Science.* V(41), 237–275. Retrieved from http://www.globalwarmingart.com/images/1/18/Arrhenius.pdf

Bakst, B. (2013, September 27). Solar garden: Model T of renewable energy? *The Christian Science Monitor.* Retrieved from http://www.csmonitor.com/Environment/Latest-News-Wires/2013/0927/Solar-garden-Model-T-of-renewable-energy

Barnidge, T. (2014, May 25). The wisdom of Mt. Diablo's solar project is reflected in its PG&E bills. *Contra Costa Times.* Retrieved from http://www.contracostatimes.com/barnidge/ci_25834608/barnidge-wisdom-mt-diablos-solar-project-is-reflected

Barwell, R. (2013). The mathematical formatting of climate change: Critical mathematics education and post-normal science. *Research in Mathematics Education. 15,*1.

Bazilian, M., & Pielke, R., Jr. (2013, Summer). Making energy access meaningful. *Issues in Science and Technology,* 74. Retrieved from http://sciencepolicy.colorado.edu/admin/publication_files/2013.22.pdf

Berbeco, M. (2013). Action in climate education: A step too far? Retrieved from Science League of America, http://ncse.com/blog/2013/12/action-climate-education-step-too-far-0015254

Berbeco, M. (2013, July 20). Teach your children: Ten things the next generation will need to know to thrive in Anthropocene. *Ensia Magazine.* Retrieved from http://ensia.com/features/teach-your-children

Berbeco, M., & McCaffrey M. (2014). Infusing climate and energy literacy throughout the curriculum: Challenges and opportunities. In J. Drake, Y. Kontar, & G. Rife (Eds.), *New trends in earth-science outreach and engagement: The nature of communication.* New York: Springer. Retrieved from http://ncse.com/files/pub/evolution/excerpt--trends.pdf

Berbeco, M., McCaffrey, M., Meikle, E., & Branch, G. (2014, April–May). Choose controversies wisely. *The Science Teacher.* 4–5.

Betts, R., Collins, M., Hemming, D. L., Jones, C. D., Lowe, J. A., & Sanderson, M. G. (2011, January 13). When could global warming reach 4°C? *Philosophical Transactions of the Royal Society A, 369*(934), 67–84.

Boden, T. A., Marland, G., & Andres, R. J. (2013). *Global, regional, and national fossil-fuel CO_2 emissions.* Retrieved from Carbon Dioxide Information Analysis Center, Oak Ridge National Laboratory, U.S. Department of Energy, Oak Ridge, TN. doi: 10.3334/CDIAC/00001_V2013

Bush, G. W. (2001, June 11). *President Bush discusses global climate change.* Presidential Speech. Retrieved from http://georgewbush-whitehouse.archives.gov/news/releases/2001/06/20010611-2.html. Also available at http://www.washingtonpost.com/wp-srv/onpolitics/transcripts/bushglobal_061101.htm

California Department of Education. (2013). *Education budget—CalEdFacts*. Retrieved from http://www.cde.ca.gov/fg/fr/eb/cefedbudget.asp

Center for Green Schools. (2013). *2013 state of our schools report*. Retrieved from http://centerforgreenschools.org/Libraries/State_of_our_Schools/2013_State_of_Our_Schools_Report_FINAL.sflb.ashx

Chakravarty, S., Chikkatur, A., de Coninck, H., Pacala, S., Socolow, R., & Tavoni, M. (2009). Sharing global CO_2 emission reductions among one billion high emitters. *Proceedings of the National Academy of Sciences, 106*(29), 11884–11888.

Cherry, L., Texley, J., & Lyons, S. (2014). *Empowering young voices for the planet*. Thousand Oaks, CA: Corwin.

Clark, D., Ranney, M., & Felipe, J. (2013). *Knowledge helps: Mechanistic information and numeric evidence as cognitive levers to overcome stasis and build public consensus on climate change*. Paper presented at the annual meeting of the Cognitive Science Society. Retrieved from http://mindmodeling.org/cogsci2013/papers/0381/index.html

CLEAN. (n.d.) *Teaching essential principle 2: Climate is regulated by complex interactions among components of the Earth system*. Retrieved from http://cleanet.org/clean/literacy/principle_2.html

Cohen, S. (2001). *States of denial: Knowing about atrocities and suffering*. New York: Blackwell Publishing.

Cohen, T., & Lovell, B. (n.d.). *Campus as a living lab: Using the built environment to revitalize college education*. SEED, American Association of Community Colleges, and Center for Green Schools. Retrieved from http://theseedcenter.org/Resources/SEED-Resources/SEED-Toolkits/Campus-as-a-Living-Lab

Colini, E., Wong, C. Y., Wilk, K. E., Curmi, P. M. G., Brumer, P., & Scholes, G. D. (2009). Coherently wired light-harvesting in photosynthetic marine algae at ambient temperature. *Nature, 463*, 644–647.

Cook, J., & Lewandowsky, S. (2011). *The debunking handbook*. St. Lucia, Australia: University of Queensland. Retrieved from http://www.skepticalscience.com/docs/Debunking_Handbook.pdf

Cook, J., Nuccitellli, D., Green, S. A., Richardson, M., Winkler, B., Painting, R., . . . Skuce, A. (2013, May 15). Quantifying the consensus on anthropogenic global warming in the scientific literature. *Environ. Res. Lett., 8*. Retrieved from http://theconsensusproject.com

Cook, J. (2014). Global warming & climate change myths. *Skeptical Science: Getting Skeptical About Global Warming Skepticism*. Retrieved from http://www.skepticalscience.com/argument.php

Crutzen, P., & Stoermer, E. (2000). The "Anthropocene." *Global Change Newsletter. 41*, 17–18.

DeWaters, J. (2009). *Energy literacy survey: A broad assessment of energy-related knowledge, attitudes and behaviors*. Potsdam, NY: Clarkson University. Retrieved from http://www.esf.edu/outreach/k12/solar/2011/documents/energy_survey_HS_v3.pdf

Diffenbaugh, N., & Field, C. (2013, August 2). Changes in ecologically critical terrestrial climate conditions. *Science, 341*(6145), 486–492. Retrieved from http://www.sciencemag.org/content/341/6145/486

Drucker, P. (1965, December 29). *The first technological revolution and its lessons*. Address to the Society for the History of Technology, San Francisco. Retrieved from http://xroads.virginia.edu/~DRBR/d_rucker5.html

Dukes, J. S. (2003). Burning buried sunshine: Human consumption of ancient solar energy. *Climatic Change, 61*(1–2), 31–44. Retrieved from http://globalecology.stanford.edu/DGE/Dukes/Dukes_ClimChange1.pdf

Environmental Protection Agency. (2006). *Life cycle assessment (LCA)*. Retrieved from http://www.epa.gov/nrmrl/std/lca/lca.html

Environmental Protection Agency. (2009). *Energy efficiency programs in K–12 schools: A guide to developing and implementing greenhouse gas reduction programs*. Retrieved from http://www.epa.gov/statelocalclimate/documents/pdf/k-12_guide.pdf

Environmental Protection Agency. (2012). *Energy use in K–12 schools*. Retrieved from http://www.energystar.gov/ia/business/downloads/datatrends/DataTrends_Schools_20121006.pdf

Figueres, C. (2013, September 27). Statement by Christiana Figueres, executive secretary, United Nations Framework Convention on Climate Change. Harvard Kennedy School of Government, Cambridge, Massachusetts. Retrieved from https://unfccc.int/files/press/statements/application/pdf/20132709_harvard.pdf

Fourier, J. (1822). *Théorie analytique de la chaleur* (in French). Paris: Firmin Didot Père et Fils.

Friedrichs, J. (2013). *The future is not what it used to be: Climate change and energy scarcity*. Cambridge, MA: The MIT Press.

Gardiner, S. (2010, January–February). Ethics and climate change: An introduction. *WIREs Climate Change, 1*, 54–66. Retrieved from http://spot.colorado.edu/~pasnau/seminar/gardiner.pdf

Gardiner, S. (2011). *A perfect moral storm: The ethical tragedy of climate change.* New York: Oxford University Press.

Global Carbon Project. (2013). *Global carbon budget: Media summary highlights.* Retrieved from http://www.globalcarbonproject.org/carbonbudget/13/hl-compact.htm

Goleman, D. (2009). *Ecological intelligence: How knowing the hidden impacts of what we buy can change everything.* New York: Crown Business.

Gross, P., Buttrey, D., Goodenough, U., Koertge, N., Lerner, L. S., Schwartz, M., & Schwartz, R. (2013). *Final evaluation of the Next Generation Science Standards.* Retrieved from http://edexcellencemedia.net/publications/2013/20130613-NGSS-Final-Review/20130612-NGSS-Final-Review.pdf

Handelsman, J., Miller, S., & Pfund, C. (2007). *Scientific teaching.* New York: W. H. Freeman.

Hassol, S. (2008, March 11). Improving how scientists communicate about climate change. *Eos, 89*(11), 106–107. Retrieved from http://climatecommunication.org/wp-content/uploads/2011/08/Eos.pdf

Hassol, S. J., & Somerville, R. C. J. (2011, October). Communicating the science of climate change. *Physics Today*, 48–53. Retrieved from http://www.climatecommunication.org/wp-content/uploads/2011/10/Somerville-Hassol-Physics-Today-2011.pdf

Hawken, P. (2009). *Commencement: Healing or stealing?* Commencement address at University of Portland. Retrieved from http://www.up.edu/commencement/default.aspx?cid=9456

Hertsgaard, M. (2012, December 5). Latinos are ready to fight climate change—Are green groups ready for them? *The Nation.* Retrieved from http://www.thenation.com/article/171617/latinos-are-ready-fight-climate-change-are-green-groups-ready-them

Hill, A. (2013, August 23). No, your phone doesn't use as much electricity as a refrigerator. *Marketplace.* Retrieved from http://www.marketplace.org/topics/sustainability/no-your-phone-doesnt-use-much-electricity-refrigerator

Holdren, J. (2014, January 8). *The polar vortex explained in 2 minutes.* The White House video. Retrieved 2014 from http://www.whitehouse.gov/photos-and-video/video/2014/01/08/polar-vortex-explained-2-minutes

Horizon Research. (2013). 2012 national survey of science and mathematics education highlights report. Retrieved from http://www.horizon-research.com/2012nssme/wp-content/uploads/2013/10/2012-NSSME-Highlights-Report.pdf

Intergovernmental Panel on Climate Change Fifth Assessment. (2013–2014). Retrieved from http://www.ipcc.ch/report/ar5

International Bureau of Weights. (n.d.). *History of the SI.* Retrieved from http://www.bipm.org/en/si/history-si

International Energy Agency. (n.d.). Retrieved from http://www.iea.org

International Organization for Standardization. (n.d.). *ISO 14000—Environmental management.* Retrieved from http://www.iso.org/iso/iso14000

Jamelske, E., Barrett, J., & Boulter, J. (2013, September). Comparing climate change awareness, perceptions, and beliefs of college students in the United States and China. *Journal of Environmental Studies and Sciences, 3*(3), 269–278.

Jamieson, D. (2014). *Reason in a dark time: Why the struggle against climate change failed—and what it means for our future.* New York: Oxford University Press.

Kiang, N. (2008, April). Timeline of photosynthesis on Earth. *Scientific American.* Retrieved from http://www.scientificamerican.com/article/timeline-of-photosynthesis-on-earth

Kolbert, E. (2014). *The sixth extinction: An unnatural history.* New York: Henry Holt.

Lackner, K. (n.d.). *Earth institute profiles.* Retrieved from http://www.earth.columbia.edu/articles/view/2523

Lashof, D. (1989). The dynamic greenhouse: Feedback processes that may influence future concentrations of atmospheric trace gases and climatic change. *Climatic* Change, *14*(3): 213–242.

Leiserowitz, A., & Smith, N. (2010). *Knowledge of climate change across global warmings' six Americas.* Report from the Yale Project on Climate Change Communication. Retrieved from http://environment.yale.edu/climate-communication/files/Knowledge_Across_Six_Americas.pdf

Leiserowitz, A., Feinberg, G., Rosenthal, S., Smith, N., Anderson A., Roser-Renouf, C., & Maibach, E. (2014). *What's in a name? Global warming vs. climate change.* Report from the Yale Project on Climate

Change Communication. Retrieved from http://environment.yale.edu/climate-communication/arti cle/global-warming-vs-climate-change

Leiserowitz, A., Maibach, E., Roser-Renouf, C. Feinberg, G., & Rosenthal, S. (2014). *Americans' actions to limit global warming in November 2013.* Report from Yale Project on Climate Change. Retrieved from http://environment.yale.edu/climate-communication/files/Behavior-November-2013.pdf

Leiserowitz, A., Maibach, E., Roser-Renouf, C., Feinberg, G., Rosenthal, S., & Marlon, J. (2014). *Climate change in the American mind: Americans' global warming beliefs and attitudes in November, 2013.* Report from the Yale Project on Climate Change. Retrieved from http://environment.yale.edu/climate-com munication/files/Climate-Beliefs-November-2013.pdf

Leiserowitz, A., Smith, N., & Marlon, J. R. (2011). *American teens' knowledge of climate change.* Report from the Yale Project on Climate Change. Retrieved from http://environment.yale.edu/climate/publica tions/american-teens-knowledge-of-climate-change

Macbeth, D. (2000, March). On an actual apparatus for conceptual change. *Science Education, 84*(2), 228–264.

McCaffrey, M. (2013, September 12). *Finding gold in the golden state.* Retrieved from NCSE blog at http://ncse.com/blog/2013/09/greetings-from-california-0015029

McCaffrey, M. (2014, May 5). *NCSE's McCaffrey on Wyoming debacle.* Retrieved from http://ncse.com/news/2014/05/ncses-mccaffrey-wyoming-debacle-0015571

McCaffrey, M., Berbeco, M., White, L., & Stuhlsatz, M. (2013, October 30). *Grappling with global change: The pedagogical challenge of the 21st century.* Presentation at the Geological Society of America, Denver, Colorado.

McCaffrey, M., & Buhr, S. (2008). Clarifying climate confusion: Addressing systemic holes, cognitive gaps, and misconceptions through climate literacy. *Physical Geography, 29*(6), 500–511.

McCright, A., & Dunlap, R. (2011). Cool dudes: The denial of climate change among conservative white males in the United States. *Global Environmental Change.* Retrieved from http://sciencepolicy.colo rado.edu/students/envs_5000/mccright_2011.pdf

Mitchell, J. M. (1976). An overview of climatic variability and its causal mechanisms. *Quaternary Research, 6,* 481–493.

Muir, J. (1911). *My First Summer in the Sierra.* Boston: Houghton Mifflin. Retrieved from http://www .gutenberg.org/files/32540/32540-h/32540-h.htm

Mundet, A. (2013, June 20). *Uncovering quantum secret in photosynthesis.* Retrieved from the ICFO-The Institute of Photonic Sciences website: http://www.eurekalert.org/pub_releases/2013-06/iiop-uqs061813.php

Munsey, C. (2008). Charting the future of undergraduate psychology. *American Psychological Association. 39,* 8. Retrieved from http://www.apa.org/monitor/2008/09/undergraduate.aspx

Nagel, T. (2014, November). *Survey analysis contradicts common climate perceptions.* Retrieved from Stanford Woods Institute for the Environment https://woods.stanford.edu/news-events/news/survey-analy sis-contradicts-common-climate-perceptions

National Academy of Sciences. (1958). *Planet Earth: The mystery with 100,000 clues.* Retrieved from http://www.nasonline.org/about-nas/history/archives/milestones-in-NAS-history/international-geophysical-year-files/igy-picture-galleries/planet-earth.html

National Center for Science Education. (2012). *Making it relevant.* Retrieved from http://ncse.com/cli mate/teaching/making-it-relevant

National Center for Science Education. (2014). *Wyoming blocks NGSS over climate.* Retrieved from http://ncse.com/news/2014/03/wyoming-blocks-ngss-over-climate-0015455

National Climate Assessment. (2014). Retrieved from http://www.globalchange.gov

National Extension Water Outreach Education. (n.d.). *Essential BEPs.* Retrieved from http://water outreach.uwex.edu/beps/essential.cfm

National Geographic Society. (2012). *Greendex.* Retrieved from http://environment.nationalgeographic .com/environment/greendex

National Geographic Society. (n.d.). *Geo-literacy initiative.* Retrieved from http://education.nationalgeo graphic.com/education/geoliteracy

National Geographic Society. (n.d.). *Network of alliance for geographic education.* Retrieved from http://edu cation.nationalgeographic.com/education/program/geography-alliances

National Oceanic and Atmospheric Administration (NOAA). (2009). *Climate literacy: The essential princi ples of climate science.* [See Appendix II].

National Research Council. (1996). *National science education standards.* Washington, DC: National Academy Press.

National Research Council. (2000). *How people learn: Brain, mind, experience, and school* (Exp. ed.). Washington, DC: National Academies Press.

National Research Council. (2001). *Climate change science: An analysis of some key questions.* Washington, DC: National Academies Press.

National Research Council. (2011). *Climate change education: Goals, audiences, and strategies: A workshop summary.* Washington, DC: National Academies Press.

National Research Council. (2012). *A framework for K–12 science education: Practices, crosscutting concepts, and core ideas.* Washington, DC: National Academies Press. Retrieved from http://www.nap.edu/openbook.php?record_id=13165&page=83

NGSS. (2012). *Next generation science standards: Frequently asked questions.* Retrieved from http://www.nextgenscience.org/frequently-asked-questions

NGSS Lead States. (2013). *Next generation science standards: For states, by states.* Washington, DC: National Academies Press.

Norgaard, K. (2011). *Living in denial: Climate change, emotions, and everyday life.* Cambridge, MA: MIT Press.

Obama, B. (2013, June 25). *Remarks by the President on climate change.* Speech at Georgetown University, Washington, DC. Retrieved from http://www.whitehouse.gov/the-press-office/2013/06/25/remarks-president-climate-change

Ohio State University. (n.d.). *Beyond weather and the water cycle.* Retrieved from http://beyondweather.ehe.osu.edu

Osborne, J. (2010). Arguing to learn in science: The role of collaborative, critical discourse. *Science Magazine, 328*(5977), 463–466.

Paramaguru, K. (2013, August 19). The battle over global warming is all in your head. *Time.* Retrieved from http://science.time.com/2013/08/19/in-denial-about-the-climate-the-psychological-battle-over-global-warming

Pew Research. (2009). *Global warming seen as a major problem around the world less concern in the U.S., China and Russia.* Retrieved from http://www.pewglobal.org/2009/12/02/global-warming-seen-as-a-major-problem-around-the-world-less-concern-in-the-us-china-and-russia

Pipher, M. (2013). *The green boat: Reviving ourselves in our capsized culture.* New York: Riverhead-Penguin.

Pipher, M. (2013, July 15). We are all climate change deniers. *Time.* Retrieved from http://ideas.time.com/2013/07/15/we-are-all-climate-change-deniers

Pugliese, A., & Ray, J. (2009, December 7). Top-emitting countries differ on climate change threat. *Gallup World.* Retrieved from http://www.gallup.com/poll/124595/top-emitting-countries-differ-climate-change-threat.aspx

Raabe, S. (2013, September 9). Xcel to triple big solar power and add wind in new generation plan. *The Denver Post.* Retrieved from http://www.denverpost.com/breakingnews/ci_24053717

Ranney, M. A., Clark, D., Reinholz, D., & Cohen, S. (2012). Improving Americans' modest global warming knowledge in the light of RTMD (reinforced theistic manifest destiny) theory. In J. van Aalst, J., K. Thompson, M. M. Jacobson, & P. Reimann, (Eds.), *The future of learning: Proceedings of the tenth international conference of the learning sciences,* (Vol. 2, pp. 2–481 to 2–482). Retrieved from the International Society of Learning Sciences at http://hamschank.com/convinceme/downloads/papers/RanneyEtAl-ICLS2012.pdf

Ray, J., & Pugliese, A. (2011, April 22). Worldwide, blame for climate change falls on humans. *Gallup World.* Retrieved from http://www.gallup.com/poll/147242/worldwide-blame-climate-change-falls-humans.aspx

Reich, C. (1970, September 26). Reflections the greening of America. *The New Yorker.* Retrieved from http://www.newyorker.com/archive/1970/09/26/1970_09_26_042_TNY_CARDS_000298460

Robson, J., Gohar, L. K., Hurley, M. D., Shine, K. P., & Wallington, T. J. (2006, May). Revised IR spectrum, radiative efficiency and global warming potential of nitrogen trifluoride. *Geophysical Research Letters, 33*(10).

Rosa, E., & Dietz, T. (2012). Human drivers of national greenhouse-gas emissions. *Nature Climate Change, 2,* 581–586. Retrieved from http://www.nature.com/nclimate/journal/v2/n8/full/nclimate1506.html

Rutherford, F., & Ahlgren, A. (1989). *Science for all Americans.* Oxford University Press.

Saylan, C., & Blumbstein, D. (2011). *The failure of environmental education (and how we can fix it)*. Berkeley, CA: University of California Press.

Shelter, M. (2013, November 21). *Americans voice strong mandate for better school buildings and infrastructure in New Nationwide independent poll*. Retrieved from U.S. Global Building Councils website at http://www.usgbc.org/articles/americans-voice-strong-mandate-better-school-buildings-and-infrastructure-new-nationwide-in

Shepardson, D., Niyogi, D., Soyoung, C., & Charusombat, U. (2011). Students' conceptions about the greenhouse effect, global warming, and climate change. *Climatic Change, 104*, 481–407. Retrieved from http://www.landsurface.org/publications-protected/J98.pdf

Sierra Club & National Council for La Raza. (2012). *National Latinos and the environment survey executive summary*. Retrieved from http://www.sierraclub.org/ecocentro/survey/2012%20Latinos%20and%20the%20Environment%20Survey_Exec%20Summary_English.pdf

Stiglitz, J., & Greenwald, B. (2014). *Creating a learning society: A new approach to growth, development, and social progress*. New York: Columbia University Press.

Strife, S., & Downey, L. (2009, March). Childhood development and access to nature: A new direction in environmental inequality research. *Organ. Environ, 22*(1), 99–122. Retrieved from http://www.ncbi.nlm.nih.gov/pmc/articles/PMC3162362/#!po=0.877193

Swann, J. P. (2009). *The 1906 Food and Drugs Act and its enforcement*. FDA History—Part I. U.S. Food and Drug Administration. Retrieved 2014 from http://www.fda.gov/AboutFDA/WhatWeDo/History/Origin/ucm054819.htm

Swim, J., Clayton, S., Doherty, T., Gifford, R. Howard, G. Reser, J., . . . Weber, E. (2009). *Psychology and global climate change: Addressing a multi-faceted phenomenon and set of challenges*. A Report by the American Psychological Association's Task Force on the Interface Between Psychology and Global Climate Change. Retrieved from http://www.apa.org/science/about/publications/climate-change.pdf

Tucker, B. (2012, Winter). The flipped classroom. *Education Next, 12*, 1. Retrieved from http://educationnext.org/the-flipped-classroom

Tyndall, J. (1861). On the absorption and radiation of heat by gases and vapours, and on the physical connexion of radiation, absorption, and conduction. *Philosophical Transactions of the Royal Society of London, 151*, 1–36.

University of Alaska at Fairbanks. (2010). *Hunting for methane with Katey Walter Anthony* [Video]. Retrieved from http://www.youtube.com/watch?v=YegdEOSQotE

United Nations Framework Convention on Climate Change. (1992). *Article 6: Education, training and public awareness*. Retrieved from http://unfccc.int/essential_background/convention/background/items/1366.php

United Nations. (2013, June 13). *World population projected to reach 9.6 billion by 2050*. Retrieved from the United Nations Department of Economic and Social Affairs website: https://www.un.org/en/development/desa/news/population/un-report-world-population-projected-to-reach-9-6-billion-by-2050.html

United States Energy Information Administration. (2012). *Energy perspectives: Fossil fuels dominate U.S. energy consumption*. Retrieved from http://www.eia.gov/todayinenergy/detail.cfm?id=9210

U.S. Department of Education. (2014). *U. S. Department of Education green ribbon schools*. Retrieved from http://www2.ed.gov/programs/green-ribbon-schools/index.html

U.S. Department of Energy. (2013). *Educational programs*. Retrieved from the National Renewable Energy Laboratory website: http://www.nrel.gov/education/educational_resources.html

Van Vuuren, D., Edmonds, J., Kainuma, M., Keyway, R., Thomson, A., Hibbard, K., . . . Rose, S. K. (2011, November). The representative concentration pathways: An overview. *Climatic Change, 109* 1–2, 5–31. Retrieved from http://link.springer.com/article/10.1007/s10584-011-0148-z

Weart, S. (2008). *The discovery of global warming* (Rev. and exp. ed.). Cambridge, MA: Harvard University Press.

Wiggins, G. & McTighe, J. (2005). *Understanding by design* (Exp. 2nd ed.). Alexandria, VA: Association for Supervision and Curriculum Development.

Wile, R. (2013, April 2). How carbon capture technology can almost turn CO_2 into cash. *Business Insider*. Retrieved from http://www.chichilnisky.com/wp-content/uploads/2013/10/Global-Thermostat-Carbon-Capture-tech-2013-4.pdf

Willard, T. (2013). *A look at the Next Generation Science Standards*. Retrieved from http://nstahosted.org/pdfs/ngss/InsideTheNGSSBox.pdf

World Bank. (2014). *Poverty overview*. Retrieved from http://www.worldbank.org/en/topic/poverty/overview

Index